SpringerBriefs in Space Life Sciences

Series Editors

Günter Ruyters
Markus Braun
Space Administration, German Aerospace Center (DLR), Bonn, Germany

More information about this series at
http://www.springer.com/series/11849

Dieter Blottner • Michele Salanova

The NeuroMuscular System: From Earth to Space Life Science

Neuromuscular Cell Signalling in Disuse and Exercise

 Springer

Dieter Blottner
Michele Salanova
Vegetative Anatomy
Charité Universitätsmedizin Berlin
Berlin
Germany

ISSN 2196-5560   ISSN 2196-5579 (electronic)
ISBN 978-3-319-12297-7   ISBN 978-3-319-12298-4 (eBook)
DOI 10.1007/978-3-319-12298-4
Springer Cham Heidelberg New York Dordrecht London

Library of Congress Control Number: 2014956534

Printed on acid-free paper

Springer is part of Springer Science+Business Media (www.springer.com)

Preface to the Series

The extraordinary conditions in space, especially microgravity, are utilized today not only for research in physical and materials sciences – they especially provide a unique tool for research in various areas of life sciences. The major goal of this research is to uncover the role of gravity with regard to the origin, evolution, and future of life, and to the development and orientation of organisms from single cells and protists up to humans. This research only became possible with the advent of manned spaceflight some 50 years ago. With the first experiment having been conducted onboard Apollo 16, the German Space Life Sciences Program celebrated its 40th anniversary in 2012 – a fitting occasion for Springer and the DLR (German Aerospace Center) to take stock of the space life sciences achievements made so far.

The DLR is the Federal Republic of Germany's National Aeronautics and Space Research Center. Its extensive research and development activities in aeronautics, space, energy, transport, and security are integrated into national and international cooperative ventures. In addition to its own research, as Germany's space agency the DLR has been charged by the federal government with the task of planning and implementing the German space program. Within the current space program, approved by the German government in November 2010, the overall goal for the life sciences section is to gain scientific knowledge and to reveal new application potentials by means of research under space conditions, especially by utilizing the microgravity environment of the International Space Station ISS.

With regard to the program's implementation, the DLR Space Administration provides the infrastructure and flight opportunities required, contracts the German space industry for the development of innovative research facilities, and provides the necessary research funding for the scientific teams at universities and other research institutes. While so-called small flight opportunities like the drop tower in Bremen, sounding rockets, and parabolic airplane flights are made available within the national program, research on the International Space Station (ISS) is implemented in the framework of Germany's participation in the ESA Microgravity Program or through bilateral cooperations with other space agencies. Free flyers such as BION or FOTON satellites are used in cooperation with Russia. The recently started utilization of Chinese spacecrafts like Shenzhou has further expanded Germany's spectrum of flight opportunities, and discussions about future cooperation on the planned Chinese Space Station are currently underway.

From the very beginning in the 1970s, Germany has been the driving force for human spaceflight as well as for related research in the life and physical sciences in Europe. It was Germany that initiated the development of Spacelab as the European contribution to the American Space Shuttle System, complemented by setting up a sound national program. And today Germany continues to be the major European contributor to the ESA programs for the ISS and its scientific utilization.

For our series, we have approached leading scientists first and foremost in Germany, but also – since science and research are international and cooperative endeavors – in other countries to provide us with their views and their summaries of the accomplishments in the various fields of space life sciences research. By presenting the current SpringerBriefs on neuromuscular physiology, we start the series with an area that is currently attracting much attention – due in no small part to health problems such as muscle atrophy and osteoporosis in our modern aging society. Overall, it is interesting to note that the psycho-physiological changes that astronauts experience during their spaceflights closely resemble those of aging people on Earth but progress at a much faster rate. Circulatory and vestibular disorders set in immediately, muscles and bones degenerate within weeks or months, and even the immune system is impaired. Thus, the aging process as well as certain diseases can be studied at an accelerated pace, yielding valuable insights for the benefit of people on Earth as well. Luckily for the astronauts: these problems slowly disappear after their return to Earth, so that their recovery processes can also be investigated, yielding additional valuable information.

Booklets on nutrition and metabolism, on the immune system, on vestibular and neuroscience, on the cardiovascular and respiratory system, and on psycho-physiological human performance will follow. This separation of human physiology and space medicine into the various research areas follows a classical division. It will certainly become evident, however, that space medicine research pursues a highly integrative approach, offering an example that should also be followed in terrestrial research. The series will eventually be rounded out by booklets on gravitational and radiation biology.

We are convinced that this series, starting with its first booklet on neuromuscular physiology in space, will find interested readers and will contribute to the goal of convincing the general public that research in space, especially in the life sciences, has been and will continue to be of concrete benefit to people on Earth.

Bonn, Germany   Prof. Dr. Günter Ruyters
September 2014   PD Dr. Markus Braun

DLR Space Administration in Bonn-Oberkassel (DLR)

The International Space Station (ISS); photo taken by an astronaut from the space shuttle *Discovery*, March 7, 2011 (NASA)

S122E008223

Extravehicular activity (EVA) of the German ESA astronaut Hans Schlegel working on the European Columbus lab of ISS, February 13, 2008 (NASA)

Preface

The term "neuromuscular system" used in the book title reflects the idea that apart from the bony skeleton, the human movement apparatus is comprised of skeletal muscles which are controlled by the neuromuscular system (i.e., brain and spinal cord sensorimotor nerve activity supports both muscle and bone quality). The main focus of this book is on recent principal findings (e.g., last 10–15 years) of the authors' major research fields in normal skeletal muscle and neuromuscular system adaptation to disuse related to Space Life Sciences by using animal research (mice and rats), human ground-based experiments (bed rest analogue to spaceflight), and the outcome of various physical countermeasures on structure, function, and cell signaling pathways of disused skeletal muscle experiments on the ground, and in actual spaceflight (International Space Station, biosatellites). This work led to some new fundamental insights, ideas, and hypotheses, and provides some future outlooks that should help to find optimized physical countermeasure protocols, e.g., more compliant to potential users and strengthening key cell signaling pathways in the neuromuscular system, e.g., nitric oxide (NO) and Homer. Some of the new insights gained from space related research might have some impact on the current understanding, in particular of human performance control and adaptation on Earth as well as in Space. We here propose a unique exercise countermeasure protocol against disuse-induced skeletal muscle atrophy and neuromuscular impairments using whole body vibration which was tested for its feasibility to prevent disuse atrophy mainly in two bed rest studies at the Charité Univertsitätsmedizin Berlin (Principal Investigator: D. Felsenberg), in cooperation with European Space Agency (ESA), German Space Agency (DLR e.V.), and industrial partners (Novotec Medical Inc., Pforzheim, Germany).

The various countermeasure protocols used in bed rest are considered as clinically controlled pioneer human studies on the ground that may help to alleviate bone and muscle loss and to support neuromuscular control to minimize risk of injury in crew members during their mission duties in Space and thereafter. We think that some of our findings from fundamental ground-based human research, for example, the anti-atrophic effects of vibration stimulation on skeletal muscle and neuromuscular key cell signaling pathways during longer periods of disuse, share the potential of being translated in routine preventive care in normal everyday life (fitness) or may be implemented as additional therapeutical tools to routine rehabilitation protocols in various clinical settings (e.g., physiotherapy, neurological or neuromuscular diseases, and osteoporosis) as well as chaperoning healthy aging on Earth.

This book is directed to all interested readers, to specialists but also to novices in the field.

Berlin, Germany

Acknowledgments

This work has been made possible only in scientific project cooperations with the help of specialized and multidisciplinary research strategies and goals and discussions with laboratories from the national and international scientific community in Space Life Science. In particular, we acknowledge our most successful cooperations in the bed rest studies with Marie-Pierre Bareille, Study Coordinator from MEDES Space Clinics, Toulouse (LTBR Study 2001, WISE Study 2005); Per Tesch from Exercise Physiology and Pharmacology, Karolinska Institutet, Stockholm, Sweden (fly-wheel technology); Alan Hargens (LBMP) from Orthopedic Surgery at UCSD, San Diego, CA, USA; Scott and Todd Trappe from Human Performance Laboratory, Ball State University, Muncie, IN, USA; Dieter Felsenberg from the Charité Universitätsmedizin Berlin, Campus Benjamin Franklin (BBR-1 and BBR-2 Studies, 2003–2006); Cecilia Gelfi and Paolo Cerretelli from the Department of Biomedical Science for Health, Segregate (Milano), Italy (BBR-1 and 2); and Jörn Rittweger from Human Physiology Department of the German Aerospace Institute (MTBR Study 2012). Our first microgravity experiment in spaceflight (MDS mission on ISS, 2009) was solicited by Vittorio Cotronei from the Italian Space Agency (ASI). The MDS project cooperation partners were Ranieri Cancedda (Principal Investigator) from University of Genova, and the MDS muscle team partners, in particular Stefano Schiaffino and Stefano Ciciliot from Venetian Institute of Molecular Medicine (VIMM), and Romeo Betto and Pompeio Volpe from Department of Neurosciences at Padova University, Italy. The cooperation partners of the second spaceflight experiment (Bion M1, 2013) were Inessa Kozlovskaya and Boris Shenkman from Institute of Biomedical Problems of the Russian Academy of Sciences, Moscow, Russia. We are also indebted to Daniel Berckmans from M3-BIORES, Catholic University of Leuven, Belgium, Marc Jamon from INSERM U 1106, Marseille, France, and Jason Hatton from ESA, Noordwijk, The Netherlands, for help with the "Mice in Space" (MIS) on ground housing tests, and Pierre Denise and Stéphane Besnard from Physiology Department of Caen University, France, for vestibular research in rodents. We also thank Lianwen Sun and Yubo Fan from Key Laboratory for Biomechanics and

Mechanobiology, School of Biological Sciences and Medical Engineering, Beihang University, Beijing, China, for ground-based rodent experiments. We greatly acknowledge support from International Space Agencies, NASA, ESA, CNES, CSA, ASI, JAXA and RSC. In particular, we acknowledge Marc Heppener from ESA-ESTEC Science and Application Division, Didier Schmitt, Head of Life Science Unit; Project managers Benny Elmann-Larsen (LTBR Study), Peter Jost (WISE Study) and Oliver Angerer (BBR-1 and BBR-2 Study) from ESA ISS Utilisation and Astronaut Support Department; Jennifer Ngo-Anh, Head of Human Research Unit at ESTEC; as well as several ESA fundings (to D.B). We are particularly indebted to Günter Ruyters, Ulrich Hoffmann, Markus Braun, and Peter Gräf, from the German Aerospace Center and Space Agency (Deutsches Zentrum für Luft- und Raumfahrt, DLR, e.V., Bonn-Oberkassel), and governmental fundings (to D.B.) from the DLR e.V. via German Ministry of Science and Education (BMBF) and German Ministry of Economics, Technology and Energy (BMWi). We also thank many co-workers from national academic institutions and departments, for example, Benedikt Schoser from Neurology Department, Friedrich Baur Institute at LMU, Munich, Dieter Felsenberg from Center of Bone and Muscle Research (CBF), Hanns-Christian Gunga (spokesman) from Berlin Center of Space Medicine (CCM), Andrew Clarke from ENT Laboratory of Vestibular Research (CBF), Sebastian Bachmann from Institute of Vegetative Anatomy (CCM), and all at Charité Universitätsmedizin Berlin, Germany. The Galileo® platform and the Galileo Space® muscle trainers are custom-made devices from Novotec Medical Inc., Pforzheim, Germany. Special thanks go to Gudrun Schiffl, Martina Gutsmann, and Petra Schrade (technicians); Jana Rudnick and Britta Püttmann (PhD candidates); Nicholas Luxa and Lisa Hilgers (MD candidates) from our laboratory; and to all other co-authors of our published original work for their excellence. Finally, our research would not have been possible without the enthusiastic help of almost 100 male and female volunteers (terrestrial astronauts) of currently five completed international bed rest studies (LTBR 2001, BBR-1 2003, WISE 2005, BBR-2 2007, MTBR 2012) and their outstanding commitment to Human Space Life Sciences.

Contents

Chapter 1
General Introduction

Abstract Skeletal muscle and the various motions performed by contracting muscle probably belong to the key ideas of animal and human life on Earth ("movement is life"). Actively contracting muscles enables the body to sustain gravitational forces which are considered as major physical impacts for development of various species and their adaptation to terrestrial life including performance control of their typical body motions and locomotion patterns, for example, during the evolutionary shift from aqueous to terrestrial life. In Space, however, human performance is challenged by gravitational unloading with loss in bone and muscle mass and impaired neuromuscular performance control comparable to those changes seen following passive motions, for example, performed by a physiotherapist, without major force production, and, more extremely, by almost complete muscle disuse (immobilization) studied in bed rest analogue studies to microgravity on Earth. To overcome disuse atrophy, reliable countermeasures need to be developed in the laboratory on the ground and in real space experiments that should help to combat disuse atrophy, but also help to stimulate functional brain areas related to sensorimotor control and to maintain neuromuscular system functions (from central spinal motoneurons to peripheral skeletal muscle targets) to support human performance control in clinical settings, rehabilitation, aging, and in extended spaceflight missions to Moon or Mars.

Keywords Human performance • Terrestrial life • Spaceflight • Disuse atrophy • Neuromuscular system • Skeletal muscle

1.1 Human Motions and Performance on Earth

Humans on Earth (earthlings) are born in gravity. The Earth's gravity (1G) is a unique environment that shapes our body and, for example, strengthens muscle and bone quality by a lifelong process of adaptation as a function of cell and tissue plasticity of many body structures including, for example, muscle and bone, cardiovascular system, and the nervous system. As our body structures and functions imperceptibly adapt to the terrestrial environment, they also help to find an adequate body equilibrium from the very first moment as a newborn or a little child, during growth and adolescence, as adults or older people, in the elderly, and finally

© The Author(s) 2015
D. Blottner, M. Salanova, *The NeuroMuscular System: From Earth to Space Life Science*, SpringerBriefs in Space Life Sciences, DOI 10.1007/978-3-319-12298-4_1

in senescence. Movement, locomotion, and performance in 1G are, however, complex, but three main principles may be critical with respect to gravitational loading adaptation on Earth:

How

- To resist gravitational forces, for example, by bone-to-bone joint lever arms to be moved by muscle contraction via force and power output so that we can move freely on the ground during voluntary movements
- To make use of the gravitational forces for local stabilization of body postures, for example, by skeletal muscle activity to enable an upright body position during standing or walking, during endurance running, or simply during sitting or prone positions, for example, at rest or sleep
- To manage adequate body equilibrium (stabilization) via neuromuscular control to perform various body motions (mobilization) which are superimposed on postural body stability as prerequisite for various kinds of motions including locomotion executed during our terrestrial life on Earth

A simplistic model of nervous system motion control in higher vertebrate evolution may be easily imagined by the conserved functional development of central sensorimotor subsystem regions in the brain and spinal cord of various species with the concomitant development of the functional locomotor apparatus comprised of bones, muscles, and peripheral nerves of the trunk and limbs by comparative functional anatomy (Fig. 1.1). In evolution of the vertebrate brain functional structures that, for example, are used in body motion control, the small brain (cerebellum) of fish comprises a relatively archetypal cerebellar region (archicerebellum) with neural networks predominantly controlling body equilibrium (vestibulocerebellum) and simple eye movements (two elementary performances of fish swimming in water). As terrestrial life evolved from aqueous life, amphibian creatures living in the transition zone between water and land were more or less challenged by altered body loading forces with the inevitable need to claim against gravity using bony structures adapted for mechanical strain compensation and contracting muscle adapted to work against gravitational forces during their terrestrial stay in absence of body lifting forces in water immersion. The amphibian motions thus evolved from a second functionally more specialized cerebellar region called the paleocerebellum (spinocerebellum) with neural networks known in higher vertebrates to be responsible for postural adjustment control and fine-tuning of purposeful motor activities via brain and spinal cord to peripheral target limb muscles (muscle tone/tension, target-directed movements) necessary for elementary body motion control (in case of amphibians by creeping and hopping on terrestrial ground). However, animal life and survival on terrestrial ground was challenged by even more complex motion patterns with the need for ready sets of well-trained motion and locomotion programs, for example, to perform flight and fight reactions, to hunt food, and to reproduce in terrestrial life. Thus reptiles, birds, and mammals including man refined (updated) their motion control system by a third functionally more specialized cerebellar region (neocerebellum) with neural networks known for the typical design of even more specialized movement

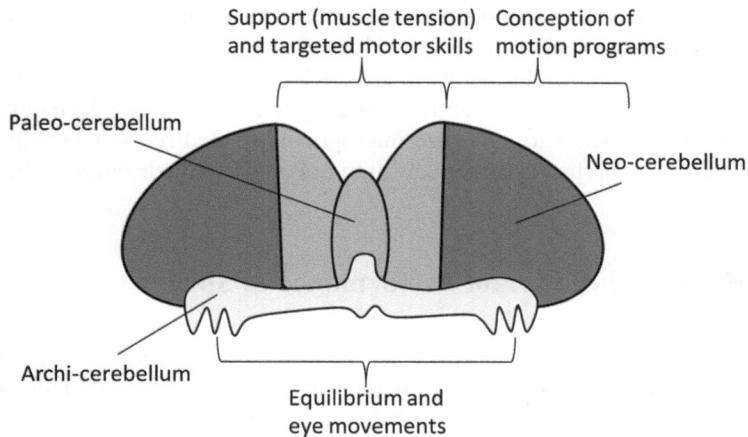

Fig. 1.1 Simplistic cartoon of the global regional assignment of the vertebrate small brain (cerebellum) during evolution. Archi-cerebellum = Vestibulocerebellum (*light*), Paleo-cerebellum = Spinocerebellum (*medium*), Neo-cerebellum = Pontocerebellum (*dark*). They represent three global regions with particular motion controls related to simple and more complex movements during evolutionary adaptation from aqueous to terrestrial live (Modified from Tews et al. 1989, p. 446)

coordination control programs for optimal survival and fitness for life (Darwinian principle) on Earth (Fig. 1.1). In all these evolutionary steps, the movement apparatus tissue structures such as bone, muscle, and nervous system turned out to share equal plasticity mechanisms for adequate adaptation to gravitational loads in terrestrial life. Though, animal and human motions performed by the musculo-skeletal system (muscle and bone) controlled by sensorimotor systems such as the neuromuscular system are far more complex than described by this simplistic view of vertebrate motion evolution. However, this example may help to imagine the impact of gravitational loading on evolution and adaptation of the body movement apparatus and motion control mechanisms on Earth. The interested reader is referred to chapters on locomotion, posture, cerebellum, and vestibular system of relevant textbooks and journal articles for more comprehensive and detailed information regarding the current understanding of motion, locomotion, and equilibrium control and the impact of gravity on Earth (Thews et al. 1989; Kandel et al. 2000; Orlovsky et al. 2001; Jamon 2014).

The term neuromuscular system used in this book is proposed in order to intuitively anticipate gravitational forces and their impact on the innervated loco-motion apparatus of human being and to find a comprehensive and more integrated view on the biomechanical and functional interactions between bone and muscle (muscle-bone unit), for example, muscle force in joint dynamic stability (An 2002) with, for example, the neuromuscular system structures (nerve-muscle unit) to enable physiological human motions and performance on Earth.

The neuromuscular system on its own and its different functional tissue and cell components are likely targets to functional adaptation on Earth but also to variable changes (deconditioning) due to the microgravity environment in Space and thus should be considered as a common functional system in the future development of inflight countermeasures to minimize microgravity-related risks of impaired performance control and health of humans in Space and after their return to Earth.

1.2 Human Motions and Performance in Space

Human motions and performance in Space is very different from terrestrial life on Earth (Wassersug 1999). In order to remain in space for longer periods such as during planned missions to Moon or Mars, the negative effects of microgravity on muscle and bone (DiPrampero et al. 2001; Cancedda 2001) and on sensimotor control, posture, locomotion, and spatial orientation (Reschke et al. 1998) must be overcome (Seibert et al. 2001).

Briefly, in weightlessness (or zero-G), the human body experiences only the acceleration that defines its inertial trajectory, or the trajectory of a free fall. In Space, almost any functional body system is challenged (deconditioned) by the zero-G or the microgravity (μG) environment. Without gravity, crew members perform their mission duties, for example, onboard the ISS, mostly in a "floating body position with minimal if any gravitational loading." For example, they receive quite unaccustomed inputs from visual, tactile, and sensory information resulting in conflicting physiological stimulation with disorientation in microgravity. The crew members in Space are also challenged by hypoactivity (relatively slow body motions that require only low forces comparable to open chain muscle activation; see Sect. 2.1.4) as their weightless body mostly takes a squatting position that may be comparable to almost floating-induced body unloading in an ambient temperature water bath on Earth which, for example, gives a pleasant feeling of wellness and full relaxation. In Space, body unloading produces disuse-induced structural, molecular, and functional changes in skeletal muscle following either short-, medium-, or long-term exposure to a microgravity environment with consequences of impaired motion and performance control. In addition to spaceflight-induced maladaptation which is called space adaptation syndrome (SAS) or space motion sickness (Lackner and Dizio 2006), the loss in bone and muscle mass and function may be termed microgravity-induced atrophy.

The impact of skeletal muscle quality on normal bone quality appears to be another very interesting aspect for future research with apparently healthy subjects to better understand the challenges of the normal human movement apparatus following extended disuse and to find optimized countermeasures for the human movement apparatus on Earth as well as in Space (Narici and de Boer 2011; Sibonga 2013). For example, bone tissue is a structure designed to resist mechanical stimuli that adapts during normal life to gain optimal biomechanical performance. In long-term spaceflight (180 days), this biomechanical homeostasis may be impaired resulting in microgravity-induced osteopenia particularly in weight-bearing bones (Vico et al. 2000; Nagaraja and Risin 2013) with a mismatch in the key functional bone cells, osteoclasts (bone resorption) versus osteoblasts (bone formation) and of related principal bone metabolism markers detectable in biological fluids (Smith et al. 2005).

In Space, loss of skeletal muscle mass and function is a well-known phenomenon, particularly in antigravity muscle such as in soleus, gastrocnemius, and other extensor muscles of the lower legs (Fitts et al. 2000). In addition, muscle protein wasting via reduced synthesis or proteolysis (Stein et al. 1999) and altered muscle-associated gene expression patterns are measurable, for example, in spaceflown murine muscle (Allen et al. 2009; Sandonà et al. 2012) as signs of muscle wasting in disuse (Bodine 2013).

A third major target of microgravity adaptation is represented by the human nervous system (Clément and Ngo-Anh 2013), in particular neural networks in the brain, brainstem, and spinal cord that control human motions and posture on Earth (Reschke et al. 1998), the sensorimotor system and reflexes (Edgerton et al. 2001), but also the neuromuscular synapses in particular of postural skeletal muscle types (Deschenes et al. 2005; Ali et al. 2009), thus compromising human motions and performance in Space (see Sect. 3.4). Space adaptation involves complex mechanisms affecting the structure, metabolism, and function of multiple body tissues and organs that may cause serious health problems both in Space and on return to Earth. Usually, the effects of microgravity are only temporarily disabling, and astronauts normally readapt to terrestrial gravitational forces quickly (Mulavara et al. 2010).

In order to minimize microgravity deconditioning and to support recovery thereafter, several physical countermeasure protocols with exercise equipment designed for use in space are available for crew members currently onboard the ISS (Macias et al. 2005; Davis and Davis 2012) or were used aboard the former Russian MIR station (Kozlovskaya and Grigoriev 2004). In addition to current available physical inflight countermeasure protocols, a set of additional countermeasures are required to adequately stimulate systemic body functions to decelerate space motion sickness, to adequately address muscle and bone wasting, and to at least partly maintain the neuromuscular control mechanisms in future extended space missions to Moon or Mars (Vernikos and Schneider 2010).

The following two chapters of this book briefly describe the current views on the structure and function of skeletal muscle and the neuromuscular system and critically discuss some of the major contributions and novel findings collected in the authors' laboratories during more than 10 years of space research at the Charité Universitätsmedizin Berlin, Germany. In particular, we here review some fundamental structural and functional changes including cellular mechanisms of the neuromuscular system in various human and animal models of simulated microgravity as well as in real space experiments. In Chap. 4, a brief outlook is given suggesting how these results may be used in future research in Space Life Sciences for the development of novel countermeasure protocols to benefit life on Earth and in Space.

References

Ali U, Fan XL, You HJ (2009) Role of muscle spindle in weightlessness-induced amyotrophia and muscle pain. Neurosci Bull 25:283–288. doi:10.1007/s12264-009-0914-3

Allen DL, Bandstra ER, Harrison BC, Thorng S, Stodieck LS, Kostenuik PJ, Morony S, Lacey DL, Hammond TG, Leinwand LL, Argraves WS, Bateman TA, Barth JL (2009) Effects of spaceflight on murine skeletal muscle gene expression. J Appl Physiol 106:582–595. doi: 10.1152/japplphysiol.90780.2008. Epub 2008 Dec 12. Erratum in: J Appl Physiol. 2011 Jan; 110(1):298

An KN (2002) Muscle force and its role in joint dynamic stability. Clin Orthop Relat Res 403(Suppl):S37–S42

Bodine SC (2013) Disuse-induced muscle wasting. Int J Biochem Cell Biol 45:2200–2208. doi:10.1016/j.biocel.2013.06.011

Cancedda R (2001) The skeletal system. In: Seibert G (ed) A world without gravity. ESA-ESTEC, Noordwijk

Clément G, Ngo-Anh JT (2013) Space physiology II: adaptation of the central nervous system to space flight – past, current, and future studies. Eur J Appl Physiol 113:1655–1672. doi:10.1007/s00421-012-2509-3

Davis SA, Davis BL (2012) Exercise equipment used in microgravity: challenges and opportunities. Curr Sports Med Rep 11:142–147. doi:10.1249/JSR.0b013e3182578fc3

Deschenes MR, Wilson MH, Kraemer WJ (2005) Neuromuscular adaptations to spaceflight are specific to postural muscles. Muscle Nerve 31:468–474

DiPrampero PE, Narici MV, Tesch PA (2001) Muscles in space. In: Seibert G (ed) A world without gravity. ESA-ESTEC, Noordwijk

Edgerton VR, McCall GE, Hodgson JA, Gotto J, Goulet C, Fleischmann K, Roy RR (2001) Sensorimotor adaptations to microgravity in humans. J Exp Biol 204(Pt 18):3217–3224, Review

Fitts RH, Romatowski JG, De La Cruz L, Widrick JJ, Desplanches D (2000) Effect of spaceflight on the maximal shortening velocity, morphology, and enzyme profile of fast- and slow-twitch skeletal muscle fibers in rhesus monkeys. J Gravit Physiol 7:S37–S38

Jamon M (2014) The development of vestibular system and related functions in mammals: impact of gravity. Front Integr Neurosci 8:11. doi:10.3389/fnint.2014.00011. eCollection 2014. Review

Kandel E, Schwartz JH, Jessel TM (eds) (2000) Principles of neural science, 4th edn. McGraw Hill, New York

Kozlovskaya IB, Grigoriev AI (2004) Russian system of countermeasures on board of the International Space Station (ISS): the first results. Acta Astronaut 55:233–237

Lackner JR, Dizio P (2006) Space motion sickness. Exp Brain Res 17:377–399

Macias BR, Groppo ER, Eastlack RK, Watenpaugh DE, Lee SM, Schneider SM, Boda WL, Smith SM, Cutuk A, Pedowitz RA, Meyer RS, Hargens AR (2005) Space exercise and earth benefits. Curr Pharm Biotechnol 6:305–317

Mulavara AP, Feiveson AH, Fiedler J, Cohen H, Peters BT, Miller C, Brady R, Bloomberg JJ (2010) Locomotor function after long-duration space flight: effects and motor learning during recovery. Exp Brain Res 202:649–659. doi:10.1007/s00221-010-2171-0

Nagaraja MP, Risin D (2013) The current state of bone loss research: data from spaceflight and microgravity simulators. J Cell Biochem 114:1001–1008. doi:10.1002/jcb.24454

Narici MV, de Boer MD (2011) Disuse of the musculo-skeletal system in space and on earth. Eur J Appl Physiol 111:403–420. doi:10.1007/s00421-010-1556-x

Orlovsky GN, Deliagina TG, Grillner S (2001) Neuronal control of Locomotion – from mollusc to man. Oxford University Press, Oxford/New York

Reschke MF, Bloomberg JJ, Harm DL, Paloski WH, Layne C, McDonald V (1998) Posture, locomotion, spatial orientation and motion sickness as a function of space flight. Brain Res Rev 28:102

Sandonà D, Desaphy JF, Camerino GM, Bianchini E, Ciciliot S, Danieli-Betto D, Dobrowolny G, Furlan S, Germinario E, Goto K, Gutsmann M, Kawano F, Nakai N, Ohira T, Ohno Y, Picard A, Salanova M, Schiffl G, Blottner D, Musarò A, Ohira Y, Betto R, Conte D, Schiaffino S (2012) Adaptation of mouse skeletal muscle to long-term microgravity in the MDS mission. PLoS One 7:e33232. doi:10.1371/journal.pone.0033232. Epub 2012 Mar 28

Seibert G et al (2001) A world without gravity – research in space for health and industrial processes. In: Fitton B, Battrick S (eds) A world without gravity – research in space for health and industrial processes. ESA Publication Division, ESTEC Noordwijk

Sibonga JD (2013) Spaceflight-induced bone loss: is there an osteoporosis risk? Curr Osteoporos Rep 11:92–98. doi:10.1007/s11914-013-0136-5

Smith SM, Wastney ME, O'Brien KO, Morukov BV, Larina IM, Abrams SA, Davis-Street JE, Oganov V, Shackelford LC (2005) Bone markers, calcium metabolism, and calcium kinetics during extended-duration space flight on the MIR space station. J Bone Miner Res 20(2): 208–218, Epub 2004 Nov 8

Stein TP (2013) Weight, muscle and bone loss during space flight: another perspective. Eur J Appl Physiol 113:2171–2181. doi:10.1007/s00421-012-2548-9, Epub 2012 Nov 29. Review

Stein TP, Leskiw MJ, Schluter MD, Donaldson MR, Larina I (1999) Protein kinetics during and after long-duration spaceflight on MIR. Am J Physiol 276:E1014–E1021

Thews G, Mutschler E, Vaupel P (1989) Funktionen des Nervensystems. In: Thews G et al (eds) Anatomie, Physiologie und Pathophysiologie des Menschen, 4th edn. Wissenschaftliche Verlagsgesellschaft, Stuttgart

Vernikos J, Schneider VS (2010) Space, gravity and the physiology of aging: parallel or convergent disciplines? A mini-review. Gerontology 56:157–166. doi:10.1159/000252852

Vico L, Collet P, Guignandon A, Lafage-Proust MH, Thomas T, Rehaillia M, Alexandre C (2000) Effects of long-term microgravity exposure on cancellous and cortical weight-bearing bones of cosmonauts. Lancet 355:1607–1611

Wassersug RJ (1999) Life without gravity. Nature 401:758

Chapter 2
Skeletal Muscle

Abstract This chapter is an update of the current view of the general structure and function of the human skeletal muscle system with its functional muscle groups (e.g., antigravity muscles) with special emphasis on the almost forgotten muscular reinforcement structures (fascia and entheses) and on the myocellular force-producing little power chambers inside skeletal muscle fibers known as sarcomeres. Skeletal muscle is well adapted to gravitational loading (1G) on Earth but is highly challenged to microgravity unloading in Space (µG, zero-G). In particular, we found that nitric oxide (NO) signaling and nitrosative stress management (via protein S-nitrosylation) in human skeletal muscle provide reliable signatures to assess efficacy of physical countermeasures. We use high-resolution confocal laser microscopy for the precise immunohistochemical biomarker detection at the cellular and subcellular level combined with quantitative biochemical methods and current protein mapping by 2-DIGE proteomics to comprehensively document disuse changes in normal healthy muscle and to assess efficacy of different exercise countermeasures against disuse atrophy. Over more than 10 years, our laboratory cooperates with national and international multidisciplinary Space Life Sciences research groups in ground-based experiments (mice) and in several bed rest studies, for example, using resistive vibration exercise (RVE) as a very efficient countermeasure against chronic disuse. Published and preliminary results from two long- and medium-term spaceflight experiments with mice, one on the International Space Station (91 days MDS mission) and another one on a Russian biosatellite on orbit (30 days BION M1 mission), provided new insights to cell signaling changes in mammalian skeletal muscle in real microgravity.

Keywords Skeletal muscle • Nitric oxide signaling • Nitrosative stress • Ground-based experiments • Physical countermeasure • Spaceflight

© The Author(s) 2015
D. Blottner, M. Salanova, *The NeuroMuscular System: From Earth to Space Life Science*, SpringerBriefs in Space Life Sciences, DOI 10.1007/978-3-319-12298-4_2

2.1 Skeletal Muscle Structure and Function

2.1.1 Functional Anatomy of Normal Human Skeletal Muscle

The human skeletal muscle system comprises about 220 specific muscles with various sizes, shapes, locations, and functions in the body. Some are relatively small (<3 cm in length) such as some hand and foot muscles (interossei, lumbricales) used for complex finger/toe movement control (grasping, playing on instruments) or some deep medial column back muscles (rotators) for local segmental spine rotations. Some are even smaller (<3 mm) and used for fine-tuning of discrete movements (Alkner and Tesch 2004) (e.g., the little stapedius controls movements of the last of three middle ear bones) in acoustic sensation and hearing. Others are longer (>40 cm, sartorius, erector spinae), while others are broader and rather powerful and fleshy (quadriceps femoris, latissimus dorsi, adductors) for use in body stabilization (posture, gait) and mobilization (movements, performance). Another functional group is located at deep muscle layers adjacent to bones running closely over one or more joints (single vs. multi-joint muscles) to facilitate local joint movement and stabilization (see Sect. 2.1.4). In general, almost 60 % of the functional skeletal musculature in the healthy human body is used for body stabilization and postural control during stance and body motions in normal everyday life on Earth and is thus termed "antigravity" muscles (Fig. 2.1).

Most of the skeletal muscles are usually grouped in the body by anatomic and functional compartments classified according to their body location and function in the trunk or extremities as ventral/dorsal, medial/lateral, and extensor/flexor compartments. For example, the soleus and gastrocnemius belong to the superficial dorsal calf plantar flexors (triceps surae) which insert via the Achilles tendon to the calcaneus bone (Fig. 2.1). The adductor muscles (short, long, and greater adductor, gracilis) belong to the medial thigh compartment (thigh adductors) for hip stabilization and leg adduction. Muscle compartments are located underneath the main body fascia (similar to a whole body stocking or cat suit) separated from the skin above (Schleip et al. 2012). The muscle compartments underneath the main body fascia particularly of the arms and legs resemble long spaces separated by fibrous connective tissue sheets known as fascia sheets or box fascia which usually contain about two or three individual muscles (e.g., the long and short fibularis of the lateral calf compartment) or two-/three-headed single muscles (e.g., biceps/triceps of ventral arm flexors/dorsal arm extensors). Each compartment is comparable to a box with functionally grouped (extensor/flexor/adductor) muscles that receive a common neurovascular supply (nerve and blood vessels) from deep main arteries or peripheral nerves and thus participate in the functional anatomy of human skeletal muscle (Blottner 2013). In most body regions, the fascia layers are part of the soft muscle tissue (Fig. 2.2) and may not be easily palpable, while others are more stiff (i.e., great thoracolumbar fascia) and well palpable on the back superficial lumbar region or at the lateral hip region (iliotibial tract) of the lower limb.

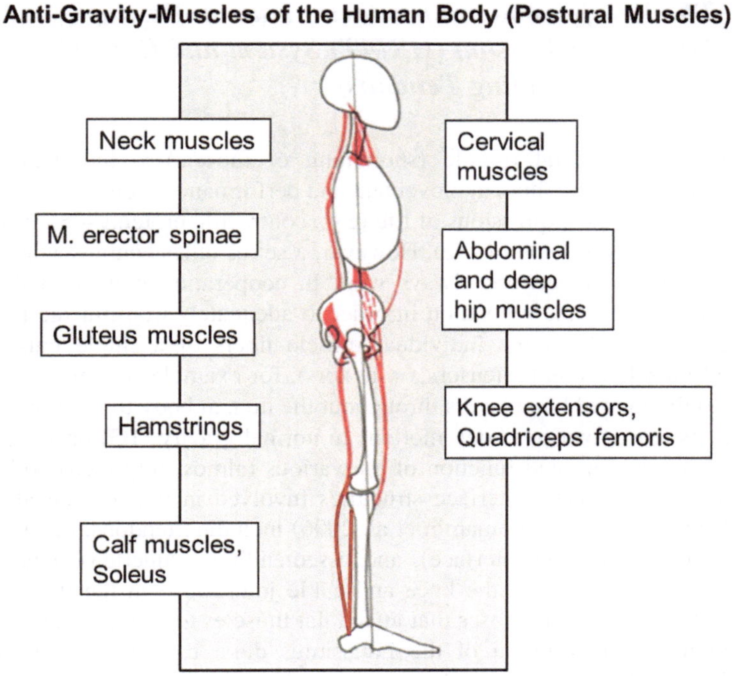

Anti-Gravity-Muscles of the Human Body (Postural Muscles)

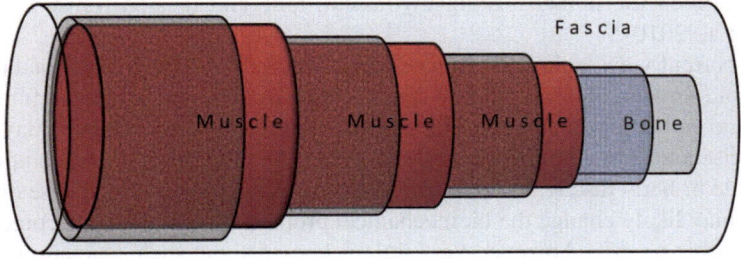

Neck muscles

Cervical muscles

M. erector spinae

Abdominal and deep hip muscles

Gluteus muscles

Hamstrings

Knee extensors, Quadriceps femoris

Calf muscles, Soleus

Fig. 2.1 Cartoon showing the key "antigravity" skeletal muscle groups of the human body for postural control of body stabilization, during stance, gait, and other motions/movements on Earth (1G)

The Myofascial Tension and Support System
Muscle, Fascia and Bone

Fascia

Muscle   Muscle   Muscle   Bone

Fig. 2.2 Proposed topographic and biodynamic model for the human myofascial tension and support system comprised of bone with periost (inner semilucent tube), up to three skeletal muscle layers (*red*) with muscle fascia (lucent tubes), and the general fascia support of the muscle compartments (Fascia, overarching lucent tube). In this model, the bone structures represent force and strain-resistant support (by relative stiffness), the muscle layers represent actively contracting elements (tension and force production), and the various fascial structures represent passive viscoelastic support (tension and force storage) by connective tissue components, collagen fiber networks, and extracellular matrix molecules. The biomechanical properties (biomechanical stress response) of active and passive structures shown here may be affected in periods of disuse (unloading) and aging on Earth and during microgravity adaptation following long-term spaceflight

2.1.2   The Reinforcement Structures: Human Resting Myofascial Tension (HRMT) System and Enthesis Organs (Including Tendons)

The contracting skeletal muscle (shortening of muscle) is the major force-producing component in human movement and performance probably representing one of the major vital expressions of life (e.g., contractile proteins $->$ muscle $->$ locomotion $=>$ expression of life). However, a set of other nonmuscular connective tissue structural network always work in cooperation with normal muscle tension, contraction, and relaxation in order to adequately transmit the local biomechanical forces from the individual muscle fibers of a single muscle of a functional muscle group (extensors vs. flexors), for example, via muscle tendons inserted to the jointed bone levers throughout the human body to reinforce muscle contractions during various body motions in normal gravity (1G) on Earth. More recently, the structure and function of the various (almost forgotten) body fascia layers and, in particular, interface structures involved in biomechanical friction termed "enthese organs" (Benjamin et al. 2006) including tendons, myotendinous junctions (muscle-tendon interface), and osseotendinous junctions (tendon-bone interface), the retinacula of the knee and ankle joint, or even the flat tendinous structures known as aponeuroses that attach flat muscles to bone, for example, the great thoracicolumbar fascia of the Latissimus dorsi back muscle, have been reconsidered as key biodynamical support structures of the human body. In addition, the dense regular and mesh-like connective tissue of a body-wide network of compression and tension (i.e., 3D tensegrity model of stiff and elastic structures) received more and more attention in the literature (Schleip et al. 2012). So far, the biomechanical roles of the body fascia under physiological and clinical aspects are only beginning to be reinvestigated as critical elements of an integrative body system termed the human resting myofascial tone and tension (HRMT) system (Masi et al. 2010).

The current view is that an estimated amount of about 10–30 % of total muscle power and force in the human body appears to be more or less stored by functional fascia and entheses related to the skeletal muscle system in normal everyday life or in exercise and sports (Huijing and Baan 2003). An altered structural composition of muscle or body fascia, for example, following longer periods of disuse or in the elderly may likely change the biomechanical properties (fascia may become either more flexible or stiff); however, very little is known about such adaptation mechanisms of the HMRT system in microgravity. As already routinely used in physical and rehabilitation medicine, physiotherapy, or osteopathy in practice, the term "myofascial tension system" should be used in addition to the term musculoskeletal system in the Human Space Life Sciences in order to underline its important structural and functional role for human performance control on Earth that were proposed to be further investigated in future ground-based analogues (bed rest disuse) and in human spaceflight (Lang T et al. Effect of long-duration spaceflight on bone and muscle: Report of the Bone-Muscle Expert Group, Cluster 1, Integrated

Systems Physiology, Theseus Project of ESA, European Science Foundation, and European Community's 7th Framework Program, FP7/2007–2013, unpublished).

In order to stimulate future research on the biomechanical properties of the various support systems of the musculoskeletal system, a comprehensive hypothetical biodynamic model for the human myofascial tension and support system is presented here that is based on functional anatomy and biodynamic properties of skeletal muscle, fascia, and bone and that is also used by the author in his academic teaching for second year medical students of the teaching module "Movement and Exercise" at the Charité Berlin, Germany (Fig. 2.2).

2.1.3   The Little Power Chambers Inside: Sarcomere, Sarcoplasmic Reticulum (SR), and More

A whole skeletal muscle is composed of larger and smaller bundles of skeletal myofibers (secondary and primary fascicles) mostly running throughout the length of a muscle and separated by perimysial connective tissue layers (Fig. 2.3). The secondary fascicles (sarcous fibers) are usually visible with the eye. In the microscope, a single muscle fiber is seen as a spindle-shaped elongated cell tube (about 20 cm in length, thickness comparable to a spiders thread, approx. 100 μm) closely packed with bundles of thin myofibrils composed of much thinner microfilaments

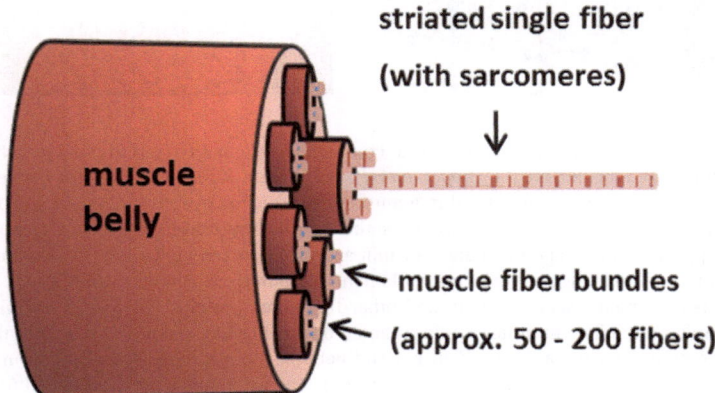

Fig. 2.3 A skeletal muscle (muscle belly) is ensheathed by a fascia and is composed of larger and smaller groups of fiber bundles (secondary and primary fascicles) embedded in loose connective tissue layers (perimysium) with neurovascular supply (capillaries, nerves, fibroblasts). A muscle fiber bundle contains approximately 50–200 striated individual fibers (polynucleated cell tube, sized 100 μm) embedded in a delicate connective tissue layer (endomysium). Each thin fiber is ensheathed by a basal lamina that provides mechanical support with extracellular matrix anchor molecules (laminin, fibronectin). Striated single muscle fibers contain bundles of myofibrils (1 μm) with their sarcomeres (little power chambers, sized 2–4 μm) in series (Fig. 2.4)

(thin actin, thick myosin filaments), with the calcium-dependent regulatory proteins (tropomyosin, troponin) responsible for actin-myosin interaction and the accessory proteins titin (spring-like protein) and nebulin (actomyosin stiffness) responsible for structural integrity, elasticity, and stiffness within the basic force-producing contractile component known as the sarcomere with its typical striated patterns (seen in light and electron microscopy) of A- and I-bands of overlapping actin-myosin microfilaments each anchored to the subtended Z-disks as the microstructural demarcations of a sarcomere (lat. *sarcos* = flesh) (Fig. 2.4, manuscript in preparation). Numerous sarcomeres (little power chambers) are arranged in series of about 500/mm of a

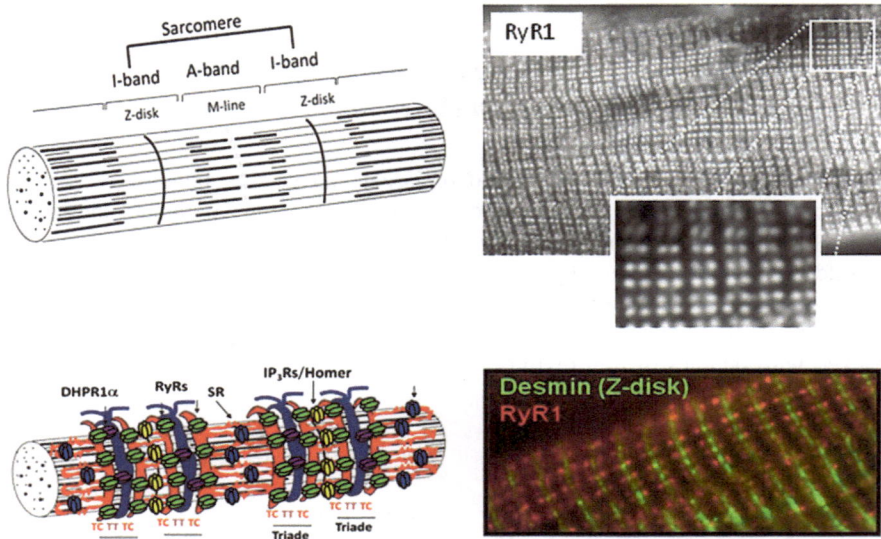

Fig. 2.4 Ultrastructural representation of a contractile sarcomere (serially power chamber) isolated from a myofibrillar bundle (*upper panel*) and its peri-sarcomeric support elements. As integral serial part of one long myofibril in a muscle fiber, the uncovered sarcomere structure has two adjacent endings (*Z-disks*) where the thin actin and thick myosin filaments are anchored and show overlapping regions. The main areas of thin actin microfilaments (*I-band*) and overlapping areas of thin actin and thick myosin microfilaments (*A-band*) are shown in the upper cartoon without support elements such as tubuli and other functional proteins. For better overview, the accessory sarcomeric filament proteins titin (filamentous spring-like stabilizers that span throughout the sarcomere from the adjacent Z-disks) and nebulin (connectivity functions during actin-myosin interactions) are not shown (*lower panel, left*). The sarcomeres are surrounded (covered) by tubular networks of the sarcoplasmic reticulum structures (SR) of the longitudinal system (L-tubuli, *thin red lines*) with terminal cisterns (*TC, thick red semicircles*) and transverse tubules (*TT, thick blue semicircles*) which are membrane extensions of the outer membrane into the muscle fiber. The main typical SR-associated calcium channel proteins are shown by four subunit protein icons of various colors, RyR1 (*green*), SERCA (*blue*), DHPR (*purple*), and IP$_3$R/Homer (*yellow*). Another protein called Homer is located adjacent to Z-disk structures but also at the subsynaptic microdomain (not shown, see Chap. 3.0). *Lower panel* (*right*) shows double staining by immunohistochemistry with anti-desmin (Z-disk marker, *green*) and anti-RyR1 (*red*) antibodies in two or three adjacent striated rat muscle fibers (Manuscript in preparation)

muscle fiber so that the microscopic shortenings of individual sarcomeres add up to the visible muscle contractions and macroscopic changes in muscle length, for example, during body motions. Skeletal muscle fibers contain all the typical cell organelles found in any other nucleated cell type (mitochondria, Golgi apparatus, endoplasmic reticulum, vesicles), but, unlike most single-nucleated cell types of the body, they contain about a hundred of myonuclei for gene transcriptional control of subcellularly partitioned microdomains for normal cell metabolism and adaptation. Muscle fibers also contain a complex and specialized endoplasmic network of tubules known as the sarcoplasmic reticulum (SR) that express a set of calcium release and calcium uptake proteins (ryanodine receptor, SERCA) which actively transport calcium ions from this major calcium store to the sarcosol and vice versa (influx/efflux) to enable normal fiber contraction during excitation. For more details, refer to a current book chapter by Brenner and Maasen (2013) and shown in this chapter (Fig. 2.18).

2.1.4   Muscle Works Against Gravitational Forces on Earth

In order to cushion the gravitational load forces, movements of the human body are biodynamically controlled via muscle by muscle contraction in accordance with a typical pattern known as "closed muscle chains" (e.g., activated in squat down by knee bending). In "open muscle chain" movements, for example, arms or legs are freely moved away from the body axis however against gravitational load but in absence of whole body weight loading (e.g., lifting of the arms or of one leg only during standing or recumbent body position). A more detailed description of the human skeletal muscle system based on topographical and functional muscle compartments is found in a current book chapter elsewhere (Blottner 2013).

In microgravity, movements of the floating human body are mainly performed according to the open muscle chain type due to gravitational unloading. However, gravitational loading can be simulated during inflight exercise by pressing the body with elastic straps fixed to comfortable body harnesses (shoulder and hip) to a fixed point of an exercise device during the training bouts, thus simulating closed muscle chain movements at least under partial resistance loading conditions in spaceflight (see Sect. 2.1.1 and next chapters).

2.1.5   Current Definitions of Human Skeletal Muscle Contraction Types Under Normal Gravity Conditions on Earth (1G)

Dynamic skeletal muscle is attached to relatively stable or stiff bone (point of origin) via more flexible osseotendinous junctions (muscle-tendon-bone) and further runs over joints to their points of insertion on another bone (usually distal to the same joint). The main function of skeletal muscle is contraction (shortening) which brings the two bones (lever arms) closer together to initiate active mode body motions during voluntary movements. This type of a "visible" muscle activity is usually considered by most people as a muscle contraction with force production. In passive mode body motions, however, muscles are passively moved (i.e., they are shortened or slightly stretched) against gravitational load (i.e., passive arm or leg movements without voluntary forces) usually performed without one's own muscle force, however, with the help of the force of a therapist (Sect. 2.2.1.1). The muscle in action is termed the agonist (e.g., biceps brachii during arm bending), while its counterpart is termed the antagonist (e.g., triceps brachii during arm bending). In this example, active and passive stretching of skeletal muscle is performed alternatingly between agonist and antagonist contractions during normal elbow joint movements.

The basic muscular contractions include **isotonic contraction** (muscle shortens in length under constant work/tension against gravity, e.g., arm lifting a heavy book from table for reading, maximal force is larger than the gravitational load of an object), **isometric contraction** (muscle remains the same length under constant work/tension against gravity, e.g., holding a heavy book in front of the body or carrying a heavy bag, the muscle force precisely matches the gravitational load), and **auxotonic contraction** (simultaneous change in length and force under constant work against gravity, e.g., heavy weight lifters, muscle force becomes higher during lifting motions). In addition, stopped preloaded contractions (e.g., masticatory pressure following teeth occlusion) and afterload contractions (e.g., lifting an object with changed effective lever arms) are composite types of contractions including isotonic, isometric, and auxotonic contractions (Brenner and Maasen 2013).

In muscle exercise, muscular strength (force) is usually increased by resistance (resistive) training which is a combination between concentric (shortening contraction, force generated while shortening occurs sufficient to overcome gravitational load) and eccentric muscle loading (lengthening contraction, force generated while muscle elongates under tension such as during deceleration of an object or lowering an object gently, the force generated becomes more and more insufficient to overcome gravitational load). Eccentric muscle contraction is typically performed during sit-ups, squatting exercises, or arm press-ups (body weight load) or with additional loading using barbells, dumbbells, or elastic straps in normal fitness or extensive strength training protocols. Compared to concentric loading, heavy eccentric loading (muscle building, strength training) has some considerable risk of muscle damage (Faulkner 2003).

2.1.6 Histologic and Molecular Adaptation in Normal, Atrophic, and Gravitational Unloaded Skeletal Muscle

Normal human skeletal muscle consists of various types of muscles with predominantly slow (type I, e.g., soleus) and predominantly fast myofiber-type (type II, e.g., brachioradialis) distribution patterns, and a variable number of muscles with mixed slow/fast fiber types (e.g., vastus lateralis) currently reviewed (Blauww et al. 2013; Schiaffino and Reggiani 2011). The normal human vastus lateralis (Fig. 2.5) of younger males (17–33 years of age, Johnson et al. 1973) consists of a mixture of fast-twitch (type II) and slow-twitch (type I) myofibers in a varying composition (approximately 65 % fast vs. 35 % slow fibers), while the antigravity calf muscle, the soleus, mainly consists of slow-type myofibers (approx. 85 % slow vs. 15 % fast). If isoform-specific MyHC subtype markers are used as detectable immunomarkers during human skeletal muscle adaptation, variable numbers of fast myofiber subtypes (IIa, IIx) are usually found, while the IIb fiber subtypes are only found in a few defined human skeletal muscle species (e.g., extraocular, pharyngeal, laryngeal). The MyHC IIb subtypes are, however, frequently found in rat or mice skeletal muscle (Ciciliot et al. 2013).

In addition to the size and phenotype distribution changes observed in disuse, the integrity of the outer muscle fiber membrane (sarcolemma) and the subsarcolemmal dystrophin scaffold structures are also compromised in human and animal skeletal muscle disuse. The dystrophin proteins are considered as key proteins of the muscle fiber subsarcolemmal scaffold helping in membrane stabilization and membrane-actin stiffness (Sarkis et al. 2013). In our work, we showed that the subsarcolemmal localized protein dystrophin also serves as a valuable immunohistochemical marker

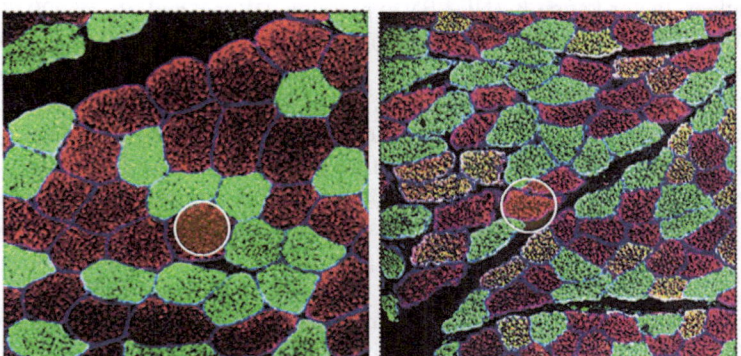

Fig. 2.5 Myofiber-type distribution (mixed 60:40 % fast type 2a,x vs. slow type 1) and cross-sectional area (*CSA*) size changes in normal (*left*) versus atrophic (*right*) human vastus lateralis (*VL*) after 60 days of disuse in bed rest (BBR-1 study, Charité Berlin). Slow type 1 (*green*), fast type 2a/x (*red*), and hybrid (*yellow*) myofibers expressing both slow and fast myosin heavy chain (s/fMyHC) immunolabels as a sign of disuse-induced fiber slow-to-fast transition. In both images, the myofiber CSA size change is marked in two cross-sectioned myofiber profiles by a white circle (approx. 4,000 mμ^2) of identical area

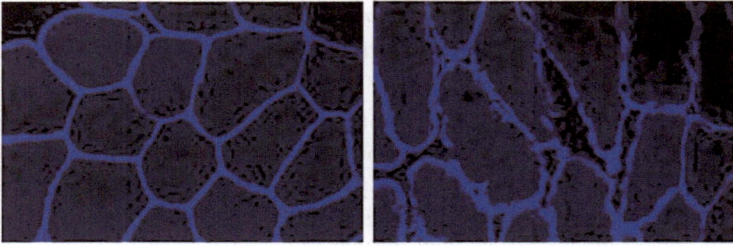

Fig. 2.6 Dystrophin immunohistochemical patterns in cross-sectioned myofibers of a normal (*left*) and chronically disused (*right*) human vastus lateralis (*VL*) reflecting disuse-induced sarcolemmal disintegration in long-term bed rest (60 days BBR-2 study, Charité Berlin, 2007)

to show altered membrane integrity patterns in normal versus disused human skeletal muscle in cryosections following prolonged disuse in bed rest (Salanova et al. 2014; Blottner et al. in press). The dystrophin immunolabel has been routinely implemented to our immunohistochemical protocols for many years now to clearly identify the outer demarcations of muscle fibers to improve qualitative and quantitative analysis of skeletal muscle fiber structures and biochemical properties on biopsy cryosections in normal versus disused human skeletal muscle as well as in space-related animal studies (Fig. 2.6).

In skeletal muscle atrophy, a fiber-type specificity exists for the regulation of muscle mass. For example, fast glycolytic fibers (type II) are more vulnerable than slow oxidative fibers (type I) that correlated with distinct signaling pathways, for example, signaling transduction of forkhead box-O family transcription factors (FOXO), autophagy inhibition, transforming growth factor-beta (TGF-ß family), and nuclear factor-kappa-ß (NF-kB) expression (Wang and Pessin 2013). The resistance of oxidative fibers to atrophy may be explained by the protective mechanisms elicited by peroxisome proliferator-activated receptor gamma coactivator 1-alpha, PGC1α (Olesen et al. 2010; Wang and Pessin 2013). A number of current reviews are available on the mechanisms modulating the muscle phenotype based on slow versus fast myofiber fiber patterns (Blaauw et al. 2013), on the molecular basis of muscle atrophy including the IGF/Akt/mTOR pathway and the myostatin/Smad pathways, on the non-lysosomal ubiquitin proteasome pathway of muscle protein degradation via the E3 ligase proteins MuRF1/Mafbx/atrogenin-1 (Jackman and Kandarian 2004; Murton et al. 2008; Bodine-Fowler et al. 1995; Bodine and Baar 2012), on the countermeasure impact on signaling pathways in disuse (Chopard et al. 2009), and on nutritional aspects and the still controversially discussed muscle protein accretion/wasting mechanisms (e.g., protein synthesis vs. breakdown) in disuse atrophy and the impact of nutrient supplementation against imbalanced muscle protein turnover in disuse (Stein and Blanc 2011) (Fig. 2.7).

Skeletal muscle disuse normally shows a slow-to-fast fiber shift likely controlled by PGC1α expression and related signal pathways upregulated by exercise (Olesen et al. 2010; Handschin 2010; Peterson et al. 2011) likely affected by physical

Anti-MuRF1 disuse atrophy   Anti-MAFbx disuse atrophy

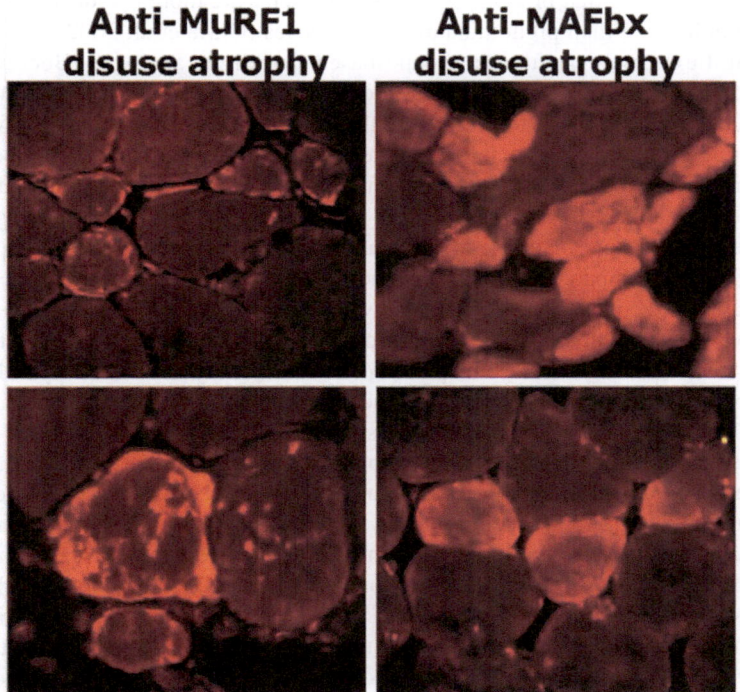

Fig. 2.7 MuRF1/MAFbx immunostaining (*red*) in cross-sectioned myofibers of disused normal human skeletal muscle fibers in female bed rest (2003 WISE-Study, Toulouse, France). Both immunolabels were used to visualize protein degradation (proteolysis) at the myofiber level following disuse (Courtesy: B. Schoser, LMU Munich (Salanova et al. 2008))

exercise as countermeasure (Desplanches 1997). An increased amount of hybrid fibers (co-expressing the slow and fast fiber phenotype, in the range of $5 < 15$ %) compared to normal ($2 < 5$ % baseline) found in various atrophy models by slow/fast myosin heavy chain (MyHC) immunohistochemistry, immunoblotting, and proteomic analysis are considered as signs of fiber-type-related transient shifting and remodeling mechanisms in various muscle disuse paradigms without and with exercise as countermeasure (Salanova et al. 2008; Salanova et al. 2009b; Moriggi et al. 2010; Luxa et al. 2013; Sun et al. 2013).

The loss in muscle mass and the basic structural and functional consequences of unloading/disuse atrophy for skeletal muscle are well described (LeBlanc et al. 1992; Fitts et al. 2001). For example, in atrophic fibers following disuse/unloading, the fiber cross-sectional area (FCSA) and the myonuclear numbers are diminished (Ohira et al. 2002), and the fiber-type distribution in a given unloaded muscle shifts from slow to fast (fiber shift or transition) resulting in reduced force-generating capacity. In gravitational unloading in animals (HU-rats), the percentage of fibers expressing fast MyHC isoforms increases in unloaded soleus but not in fast muscles suggesting that predominantly slow muscles are more responsive to gravitational unloading than predominantly fast muscles (Ohira et al. 2002).

Even if extremely useful as a ground-based spaceflight analogue (Booth 1994), disuse in bed rest may be an extreme model of physical inactivity (Thyfault and Booth 2011), for example, to investigate the changes following sedentary lifestyles or even aging of modern societies. However, bed rest immobilization appears to also mimic many of the structural and functional changes observed in sedentary lifestyle residents (loss in muscle mass and force, others) either challenged by extended sitting time or much to low daily stepping patterns (<3,000 steps/day/ healthy adults) compared to a physically active lifestyle with approximately > 7.000 steps/day of a recommended physical activity (Tudor-Locke et al. 2011).

However, the actual fiber composition in a normal healthy skeletal muscle largely depends on genetic predispositions and may vary due to lifestyle, history of activity, and sex and age of a subject (Baldwin and Haddad 2002). After analyzing more than 800 muscle biopsies from mainly slow and slow-/fast-mixed muscle types of healthy human subjects over the last decade, we suggest that apart from the known intersubject variability and sex differences, for example, between male and female muscle, a continuum of the myofiber-type distribution should be considered because of ongoing adaptation processes throughout normal human life on Earth. In laboratory rats and mice of comparable genetic strains, the muscle fiber size and type changes are "more stable" due to controlled housing conditions in standard vivarium cages (IVC cage) with defined group sizes, activity status, and feeding of standardized nutrients. Even smaller cages (<50 % reduced floor area vs. IVC cages) may not confound structural biochemical and behavioral analysis of laboratory mice (C57Bl/6 strain) even under fully automated life support housing conditions usually required for rodent spaceflight experiments (Blottner et al. 2009).

Due to the relatively well-known fiber-type composition of human skeletal muscle and due to their well-palpable anatomical location in the human body, two of them, the vastus lateralis and the soleus, have been considered by many laboratories around the world as standard reference muscles for the reliable statistical analysis in previous bed rest studies and most probably also in crew members for comparability reasons of ground and flight sample analysis.

2.1.7   NO Signals in Muscle: Nitric Oxide (NO) Produced by NO-Synthase (NOS) as Multifunctional Signals in Normal Muscle Physiology

With the discovery of nitric oxide (NO), formerly known as endothelial-derived relaxation factor (EDRF), biological gas signals were of particular interest in fundamental muscle biology and physiology (Bredt and Snyder 1990), disease (Brenman et al. 1995), clinical applications (Zhou and Zhu 2009), and, more recently, also in sports and exercise biology (Suhr et al. 2013). Biological NO signals were described in the late 1980s (around 1987–1989) as a completely new class of diffusible gaseous signals first described in vascular endothelial cells (NO as vasodilator) and in subsets of neuronal cells (NO as diffusible neurotransmitter, long-term potentiation) of the brain by several scientists including Salvador Moncada from the UK, but also by the US scientists Louis Ignarro, Ferid Murad, and Robert Furchgott (Nobel Prize in physiology and medicine in 1998). In addition to NO signals found in the vasculature (endothelial NOS, eNOS) and central and peripheral neuronal networks (neuronal NOS, nNOS), skeletal muscle soon was identified as another large source of NO signals in the body (Nakane et al. 1993; Silvagno et al. 1996) including ventilatory muscles (Hussain et al. 1997). The skeletal muscle-derived NO signals and their dynamic roles in contraction and force mechanisms, myocellular metabolism, and cell signaling pathways became of particular interest for skeletal muscle biology and various adaptation mechanisms related to muscle structure and physiology. Currently, at least three individual NO-synthase genes are known, termed NOS1 (nNOS, neuronal), NOS2 (iNOS, inducible), and NOS3 (eNOS, endothelial) according to the proposed new terminology for NOS isoforms (Stamler and Meissner 2001) with about 20 different splice variants (Brenman et al. 1997) (Fig. 2.8). A muscle-specific splice variant of NOS1, termed mμNOS (Silvagno et al. 1996), was characterized as part of the dystrophin-glycoprotein complex of the skeletal muscle fiber membrane (sarcolemma) where mμNOS/NOS1 is anchored through a PDZ motif at its NH2 terminus to alpha and beta syntrophins and thus to dystrophin and dystrobrevin (Grozdanovic and Baumgarten 1999; Percival et al. 2008).

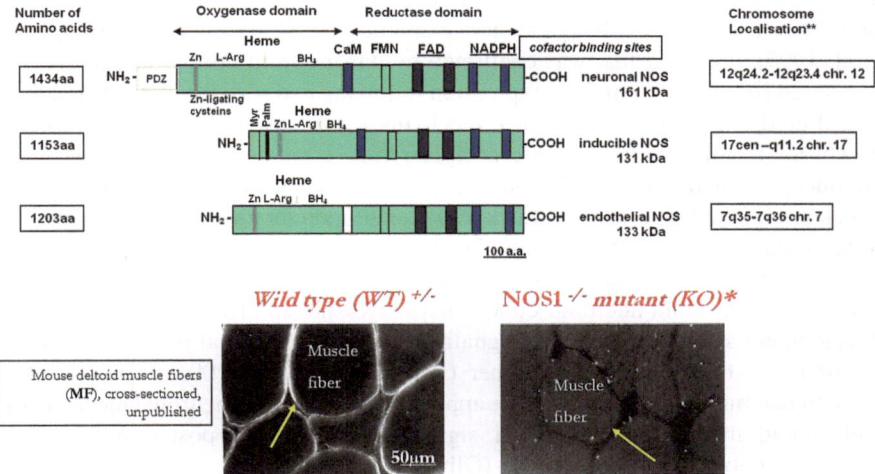

Fig. 2.8 Domain structure of NOS 1–3 genes on three human chromosomes. Skeletal muscle sarcolemmal nNOS immunolocalization (*arrows*) in wild-type versus nNOS$^{-/-}$ mutant mice deltoid muscle fibers (cross sections), D. Blottner, unpublished. Cartoon (*upper panel*) mod. from Ignarro LJ (ed) Nitric oxide biology and pathobiology, Academic Press, p.12 (2000)

NO signals are considered as metabolic regulators during muscle exercise in health and disease (Kingwell 2000) but also during muscle denervation and reinnervation (Tews et al. 1997; Rubinstein et al. 1998) and in muscle repair and remodeling mechanisms via satellite cells (Anderson 2000). Imbalance in NO signal mechanisms results in movement impairments such as slowing the walking speed in rats following in vivo NOS inhibition (Wang et al. 2001). Our previous basic research on the muscular NOS/NO system aimed at investigating the spatio-temporal expression and localization patterns using immunofluorescence methods with isoform-specific NOS antibodies in conjunction with nonradioactive NOS mRNA in situ hybridization techniques in skeletal muscle development that showed that all three major NOS1-3 isoforms are basically expressed in mouse C2C12 single myoblasts that are transiently upregulated when the single-nucleated myoblast fusion to become elongated multinucleated myotubes in vitro that eventually differentiate into mature skeletal muscle fibers (Blottner et al. 1998). Other work showed that explanted single muscle fibers are able to release measureable NO signals into the culture medium upon in vitro drug administration (NO-donors, arginine, or nitroprusside, Balon and Nadler 1994) and following in vivo electrical stimulation in the rabbit possibly involved in metabolic control of fiber phenotype transformation of anterior tibialis and extensor digitorum muscle (Reiser et al. 1997). In addition, mechanical loading was found to affect NOS expression by in vitro (myotubes) and in vivo (HU-rat) experiments confirming activity-dependent muscular NO signaling (Tidball et al. 1998; Fujii et al. 1998). Other studies in adult animals found that the sarcolemmal NOS is able to translocate from membrane to the sarcosol (cytosol) in response to mechanical unloading (Suzuki et al. 2007). Interestingly, the NOS2 isoform (inducible NOS) was found to be associated with the outer membrane protein caveolin-3 in slow-twitch guinea pig muscle suggesting involvement, for example, in glucose metabolism (Gath et al. 1999), a finding that was confirmed by our group in normal human muscle by altered NOS2-caveolin-3 co-expression following bed rest (Rudnick et al. 2004). In cultured C2 myotubes (mouse C2 cell line), we showed that NOS1 proteins co-cluster together with nicotinic acetylcholine receptors (major component of cholinergic synapses) by agrin-induced postsynaptic membrane clusters in vitro (Lück et al. 2000; Blottner and Lück 2001), and upon motor nerve contacting, the myocellular NO signals were upregulated in C2 myotubes in nerve-muscle cocultures performed in our laboratory (Püttmann et al. 2005). Nitric oxide synthase is expressed at the adult neuromuscular synapse (Kusner and Kaminski 1996) during development suggesting multiple signaling functions for neural transmission from motor nerve to skeletal muscle fiber (Blottner and Lück 2001). These studies showed that muscular NO signals are apparently regulated by neuronal mechanisms and should therefore have crucial signaling functions in postsynaptic neuro-muscular control of skeletal muscle (Oliver et al. 1996).

Earlier work from our laboratory also suggested a sarcolemmal NOS1 localiza-tion as shown by NOS immunohistochemistry and also by the NADPH diaphorase enzyme histochemical activity (histochemical marker of NOS activity) in develop-ing rat and mouse normal skeletal muscle by our group (Blottner and Lück 1998;

Lück et al. 1998). In those studies, we found a gradient of increased sarcolemmal NOS expression in developing mice muscle fibers before and around birth (perinatally). In further animal and human studies, investigations on the muscular NOS/NO system turned out to serve as a valuable biomarker to study disuse atrophy (loss and/or translocation of myocellular NOS) and to assess efficacy of exercise countermeasure (preservation of myocellular NOS) in bed rest (see Sect. 2.2.3.4).

2.2 Lessons Learned from Ground-Based and Spaceflight Experiments

2.2.1 Animal Studies on Earth

Up to now, established and current in vitro models use isolated cells and tissues and multipotent stem cells cultivated in defined nutrient solutions under controlled conditions in special laboratory incubators. They are useful tools to study fundamental mechanisms in cell biology and physiology including cell development, differentiation, and growth, for example, triggered by bioactive molecules (growth factors, cytokines) by pharmacological in vitro cell assays as standard methods routinely used in research laboratories around the world. In gravitational and radiation biology, isolated primary cells, cell lines, or tissues are used to study microgravity changes in plant and animal cell biology grown in special cell incubators as payload to study fundamental mechanisms "unmasked of gravity" (Unsworth and Lelkes 1998), for example, cell shape and cytoskeleton changes in gravisensing (Hughes-Fulford 2003), bone cell signaling (Hughes-Fulford 2004), and cartilage differentiation (Duke and Montufar-Solis 1999), just to mention some of the cell biology experiments already flown in Space.

Skeletal muscle research offers a battery of myogenic cells (myoblasts, myotubes), isolated muscle fibers, and muscular stem cells (satellite cells) that are currently available to study defined aspects of physiological and pathophysiological mechanisms (e.g., fiber contractility, muscle development, and regeneration on Earth (Aas et al. 2013) as well as in simulated microgravity (Benavides Damm et al. 2013) and in spaceflight experiments (Franzoso et al. 2009).

However, isolated single cells or cell layers from various tissues sources (bone, muscle, brain, others) may not fully replace in vivo studies in small laboratory animals (rats and mice) as, for example, currently demonstrated by a successful stem cell therapy (stem cells were cultivated in vitro and afterward implanted in vivo) to offset skeletal muscle atrophy in mice (Artioli et al. 2014).

In vivo studies are essential to further understand the complexity of integrative signaling processes and their adaptation mechanisms inherent to the neuromuscular system in a normal living higher vertebrate organism including also humans. Due to cost limitations and other spaceflight constraints (design of animal housing devices for spaceflight), most of the animal studies in space research are performed by international scientific laboratories on the ground with expert staff and adequate equipment in animal facilities according to commonly accepted regulations for use and care of animals in scientific research consented and approved by national and local ethical board members of various disciplines (NRC 1996; Katahira 2001).

2.2.1.1 Ground-Based Experiments I: Hind Limb Unloading (HU) in Rats

Hind-limb unloading (HU) of the rat is a traditional and well-established ground-based experimental animal research paradigm in spaceflight research used by NASA to study disuse atrophy in small rodents such as rats (Morey-Holton and Globus 2002) or mice (Ferreira et al. 2011). Experiments with the HU-rat model require specially designed cages as animals need to move freely with both fore paws touching the ground for water and food ad libitum while the rat's body is in a 30° angle to secure for hind-limb unloading (HU position, i.e., tail suspension) with both hind limbs freely hanging down without touching the bottom of the cage. Due to the common regulations of animal use and care in research, only laboratories with specialized animal facilities are able to use rats or mice over several days or weeks for HU experiments (Morey-Holton and Globus 2002). Because of ethical reasons, HU animal studies should be performed in cooperation between laboratories for optimal tissue sharing and analysis (Morey-Holton et al. 2007).

In a bilateral and interdisciplinary university cooperative between the Charité Berlin, Germany, and the Beihang University Beijing, China, we were able to share muscle tissue from a rat 21d HU experiment performed with a custom-made stepper device for rodents by our Chinese partners from the Key Laboratory of Biomechanics and Mechanobiology at the Beihang University to study bone and muscle loss under active versus passive muscle training (Sun et al. 2013). We found that the NOS1 biomarker in HU-rat soleus muscle almost disappeared from the unloaded myofiber sarcolemma (after 21 days of HU) compared to control animals (vivarium control). This loss in sarcolemmal NOS 1 was, however, prevented by active stepper training of hind limbs during HU position (reflexive muscle contractions triggered by plantar electric pulse arc stimulation) and to lesser extent also by passive stepper training during HU (passive mode motions induced by up and down movements of hind legs during HU via motorized foot pedals) compared to relevant controls performing comparable stepper training in normal body position in vivarium standard cages (Sun et al. 2013). This direct comparison between active and passive motions of disused rat skeletal muscle can be experimentally investigated only by the HU protocol that provided conclusive evidence for the presence of an activity-driven sarcolemmal NOS1 translocation from the myofiber sarcolemma to the sarcosol that can be offset by adequate muscle training (Sun et al. 2013). The results are interesting because activity-driven NO signals control a set of functional muscle-specific proteins via protein-S-nitrosylation, and, for example, abundant NO signals (too much or too little) could explain some of the altered muscular functions observed in human skeletal muscle disuse by nitrosative stress mechanisms experimentally proven in bed rest by our laboratory (see Sect. 2.2.3.5). In the same HU experiment, active stepping of gravitationally unloaded rat hind limbs prevented loss in tibial bone mass (the soleus originates from to the tibia) and trabecular microarchitecture, supporting the notion for optimization of human countermeasure protocols that should be targeted to individual human muscle-bone units following disuse on Earth, in rehabilitation, or in spaceflight.

2.2.1.2 Ground-Based Experiments II: Vestibular Deafferentation in Rats

As already discussed in the introduction chapter, the vestibular system (organ of equilibrium) is a key equilibrium and spatial reference system for adequate body movement control on Earth that undergoes considerable modifications during spaceflight with obvious consequences on spatial orientation and well-being in crew members who, for example, may suffer from space motion sickness (Lackner and Dizio 2006). Previous findings from the literature suggested that the vestibular outflow may control general sympathetic outflow in the body and, in particular, also the muscular sympathetic nerve activity on Earth as well as in Space. For example, baseline sympathetic outflow was found to be increased as a reason for hemodynamic stress in humans in Space (Eckberg et al. 2003); human sympathetic nerve activity is reduced after short microgravity exposition (parabolic flight) and upregulated after longer periods of microgravity (spaceflight) with impaired arterial baroreflex (Mano and Iwase 2003; Mano 2005) concomitant to altered human muscle sympathetic nerve activity (Ertl et al. 2002). However, the obvious role of vestibular changes for skeletal muscle disuse atrophy received only little attention (Kasri et al. 2004; Tothova et al. 2006).

In a laboratory cooperation between the Physiology Department at the University of Caen, France, and the Neuromuscular Group at the Charité Berlin, we tested the hypothesis that the vestibular system may affect both myofiber size and type composition of skeletal muscle in normal adult rats ($n = 8$) subjected to vestibular deafferentation (labyrinthectomy) in a 30-day long laboratory experiment established by our French partner laboratory. This experiment made use of surgical removal of the vestibular system in rats or mice to extinct the gravity stimulus by destruction of the gravisensor on Earth (Jamon 2014). Compared to sham-treated controls (with still intact gravisensors), in vivo bilaterally vestibular lesioned rats now missing the graviceptors showed significant myofiber size reductions in tissue samples of postural soleus muscle, altered fiber-type composition, and reduced myonuclear NFATc1 transcription factor accumulation (i.e., NFATc1 shift from cytosol to nucleus), all signs of slow myofiber atrophy remodeling and transcriptional activity changes in slow myofibers of postural soleus muscle (Luxa et al. 2013). Interestingly, vestibular lesions also resulted in considerable bone loss in this animal model (Denise et al. 2009) with clinical implications considering the identification of a unique inner ear signal control of bone formation (Vignaux et al. 2013) and possibly also acting on skeletal muscle quality (Luxa et al. 2013).

Even if the exact neuroanatomical links remain to be elucidated, it seems likely that, similar to bone formation, skeletal muscle adaptation may be controlled by autonomic nervous system signals via spinal vestibulosympathetic outflow to peripheral targets to support normal skeletal muscle mass and function under terrestrial gravity (Luxa et al. 2013). These findings are of particular interest to find more adequate countermeasures against microgravity-induced muscle atrophy and bone loss for the crew members in future spaceflights. These findings also may have some impact on the clinical management of common vestibulosympathetic disorders (Ménière's disease), for example, for better interpretation of clinically feasible skeletal muscle monitoring by vestibular evoked myogenic potentials to reliably determine inner ear vestibular saccule function (Honaker and Samy 2007).

2.2.2 Animal Studies in Spaceflight

Since the early days of spaceflight, animals were frequently flown in Space by NASA and the Russian Space Agencies onboard satellites on orbit (Cosmos) for several days up to a few weeks (Morey-Holton et al. 2007; Ballard and Rossberg Walker 1992; Oganov et al. 2006). A review on the history of spaceflown animals "from the first dog to the last monkey in space" can be found elsewhere (Ilyin 2007). Animal experiments in space with higher vertebrates and mammals are essential to study the physiological mechanisms working in the human body in microgravity (Yamasaki and Shimizu 2004) in sufficient statistical numbers. Our group participated to an international muscle team involved in two flight experiments with rodents in Space, one in 2009 (MDS mission, Sect. 2.2.2.1) with mice housed for over 90 days in a special animal payload on the ISS (mice drawer system, MDS, 2009), supported by the Italian Space Agency (ASI), NASA, and JAXA, and another one in 2013 with adult mice flown on a Russian biosatellite (BION-M1, Sect. 2.2.2.2), supported by the DLR (Germany) and IMBP (Russia).

It should be noted that mice are housed in Space in smaller cages (reduced area cages) than normally used in animal vivarium on the ground (standard ICV cages) with the potential risk of confounding results obtained by spaceflight experiments with mice living on reduced area (floor size) with, for example, less activity levels and challenged by higher stress levels (noninvasively analyzed by corticosteroid monitoring of feces). However, pilot laboratory experiments with C57/Bl6 mice housed in automated life support cages with reduced floor areas (MRSM cages, Fig. 2.9) for 3 weeks at the laboratory of our cooperation partners of the Measure, Model, and Manage Bio-Responses (M3-BIORES) working group at the University of Leuven (Kastelpark Arenberg, Heverlee), Belgium, revealed little if any negative effects on musculoskeletal system and behavioral outcome (stress and learning) produced by the unique cage design (Fig. 2.9) suggesting that the new environmental condition may not confound structure, physiology, and behavior results from Space Life Sciences research experiments and suggesting that mice pre-adapted to fully automated life support animal habitats for about 1 week prior to

Fig. 2.9 Ground-based mouse science reference module (MRSM) experiment performed together with an ESA/ASI supported multidisciplinary MISS Facility Science Team (FST) of scientific experts (L. Vico, M. Jamon, D. Berckmans, D. Blottner), ESA representatives (P. Schiller, O. Angerer, J. Hatton, F. Gaubert), ASI representatives (V. Cotronei), and industrial partners from Thales-Alenia Space (G. Falcetti) located at the Catholic University of Leuven, Belgium, in 2006–2007. Biological testing was performed for any structural, physiological, and behavioral parameter changes of male C57Bl/6 mice housed (singly or in pairs) in small animal habitats (cf. three reduced floor size cages seen in middle racks) compared to standard-sized IVC cages integrated in the same rack shown (at the right side). Each mice habitat (cage) has an automated food supply system on its rear wall, a waste filter system underneath the cage bottom. Cages are air-conditioned through a closed tube system connected to an automated life support system (not shown) for ambient air ventilation, temperature, and humidity control (Blottner et al. 2009)

launch (astromice) should be implemented in future scenarios of spaceflight experiments (Blottner et al. 2009).

2.2.2.1   Mice on the International Space Station (ISS). MDS Mission

In 2009, a spaceflight payload for rodent research on the ISS, the mice drawer system (MDS), was initiated by the Italian Space Agency (ASI) and designed by Thales Alenia Space Italia company (Fig. 2.10). The MDS payload was flown to the ISS via the Shuttle Discovery 17A/STS-128 (Aug 28th, 2009) and returned to Earth with Shuttle Atlantis ULF3/STS-129 (Nov 27th) after 91 days in Space, which currently still is the longest permanence of mice in Space ever (Cancedda et al. 2012). The MDS housed six mice (3 wild-type and 3 PTN-Tg mutant mice, overexpressing a bone-specific promoter for pleiotropin (Fig. 2.10). Unfortunately, three mice died during the MDS mission due to health status and payload-related reasons. The remaining mice showed a normal behavior during the experiment, food and water and health status were daily checked, and they appeared in excellent

Fig. 2.10 The 91 days duration mice drawer system (MDS) module scenario (Aug 28 to Nov 27, 2009, STS 128/129, PI Cancedda) on the International Space Station (ISS). NASA Astronaut Nicole Scott onboard the ISS at the Japanese Kibo Laboratory with the MDS module (right side) housing three (n = 3) C57Bl/10 J and three (n = 3) osf-1 transgenic mice (Courtesy: NASA (http:// onorbit.com/node/1601))

health after landing. To receive as much information as possible on the microgravity-induced changes, spaceflown mice were immediately sacrificed upon return from Space, the various tissues of interest dissected within 24 h after landing and distributed to a Tissue Sharing Team of about 20 research teams from 6 countries including Germany (Charité Berlin). A ground replica of the flight experiment including animal housing in a second ground-based MDS drawer device was performed in parallel to the spaceflight experiment in the PI's laboratory (R. Cancedda) of the University of Genova, Italy. In addition, control tissue was also obtained from mice housed in standard vivarium cages on Earth (Cancedda et al. 2012).

The skeletal muscle changes of the MDS mice include a slow-to-fast transition of the soleus myofibers (type 1/type 2 shift) and slow/fast myosin heavy chain content typically for this postural muscle following extended microgravity unloading. The sarcolemmal NOS1 translocation to the sarcosol was also found after prolonged microgravity exposure (Sandona et al. 2012) as well as flight-related NOS1–3 gene expression in a muscle-specific pattern (SOL vs. EDL) by microarray analysis (Fig. 2.11) confirming the previous ground-based experimental results reported by our group exposed to simulated microgravity models in animals (HU) and humans (bed rest). Moreover, a set of atrophy-related ubiquitin ligases (MuRF E3 ligase), sarcolemmal ion channels (NAV1.4, K + −channel subunits Kir6.2, SUR2A, and SUR1), various stress-related genes (NF-kB), and transcription factors (MRF-4) were upregulated in spaceflown soleus muscle compared to ground

Fig. 2.11 Nitric oxide synthase isoform (NOS1-3) gene expression (microarrays) in mouse soleus (*SOL*) and extensor digitorum (*EDL*) skeletal muscle following long-term spaceflight (flight) onboard the ISS (91d MDS mission) compared to ground controls (ground) (Manuscript in preparation)

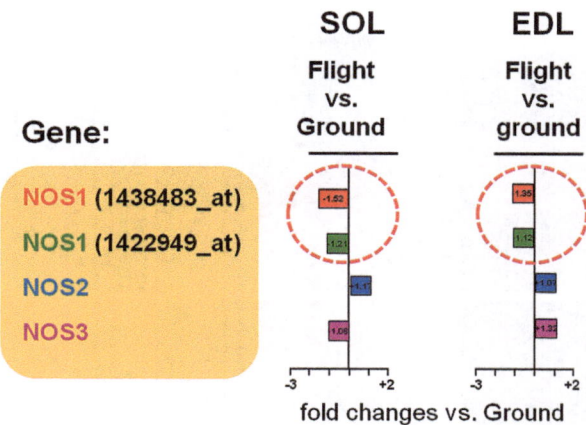

controls (Sandona et al. 2012). In conclusion, antigravity muscles such as the soleus react in a very sensitive way against prolonged microgravity exposure, while other calf muscles (EDL) were more resistant to unloading probably by activating compensatory and protective signaling pathways. These results support the idea to identify molecular targets for the new development of countermeasures (Sandona et al. 2012). Some of the results of the MDS mission, for example, on bone, muscle, and brain of the spaceflown mice including our data on skeletal muscle, are published elsewhere (Cancedda et al. 2012; Ohira et al. 2014; Santucci et al. 2012; Tavella et al. 2012; Sandona et al. 2012; Camerino et al. 2013).

2.2.2.2 Mice Onboard a Biosatellite: 30 Days BION-M1 Flight Campaign

The BION satellite program for investigations of microgravity changes on various biological organisms and organ systems of higher vertebrates has a long tradition in Russian spaceflight research (Ilyin 2000). In a university cooperation between the Charité Berlin, Germany, and the Institute of Biomedical Problems (IMBP) of the Russian Academy of Sciences, Moscow, Russia (kindly supported by DLR/BMWi grants), our laboratory had the exciting possibility of participating to a 30-day BION-M1 flight campaign with young adult male mice housed separately by groups of six animals per cage in a controlled animal life support device with automated food and water supply on a biosatellite flown to orbit by a Soyuz-2 rocket (Andreev-Andrievskiy et al. 2014). The flown mice that returned back to Earth in life and in good health were sacrificed 24 h after landing by a dissection team at IMBP, and various tissues were distributed to each laboratory according to a special tissue preservation plan (sample preparation) and according to a dissection schedule consented by all investigators prior to end of the 2013 BION M1 mission (Fig. 2.12). Our laboratory received samples from six different skeletal muscles

Fig. 2.12 The BION-M1 biosatellite flight experiment scenario (April 19 to May 19, 2013). The BION M1 biosatellite with mice and other biological samples was flown by SOYUZ-2 rocket to near orbit (575 km altitude, x477 orbital circulations) and returned to Earth (Kasachstan, Russia) after 30 days (*upper left*) with soft landing system. Flown mice were harvested on landing site and delivered to the IBMP, Moscow, 11 h post landing (*upper right*). Three C57Bl/6 SPF mice (aged 4–5 months, approx. 28 g b.wt. each) were housed in a small container (lower left showing 4 of 5 module containers) with food and water ad libitum, 12 h light/dark cycle, and under ambient environmental parameters (21 °C temperature, 60 % rel. humidity, pO_2 140–180 mmHg concentration, pCO_2 traces). Mice were sacrificed, and tissue samples prepared by a well-experienced team of scientists (*lower right*), frozen fixed, and delivered to the principal investigator's laboratories (Images: IMBP, Moscow, Russian Federation)

from five spaceflown mice each (n = 5, BION flight) and from a number of age- and sex-matched mice from three control groups (BION ground, flight control, vivarium control, each n = 6 or n = 8) that are currently being analyzed.

Preliminary results obtained from first analysis of the BION samples confirmed that the postural soleus and the longissimus dorsi back muscle of flown mice are highly atrophic compared to ground-based controls (ms. in preparation). In addition, intensity of the NOS1 immunoreactivity is changed in the soleus muscle compared to ground controls confirming the presence of microgravity-induced control mechanisms related to the muscular NOS/NO system in mice previously found in HU-unloaded rats (Sun et al. 2013) but also in disuse atrophy in bed rest (Rudnick et al. 2004). Further analysis, for example, on the transcriptional level, gene expression, and various putative muscular signaling pathways involved in gravitational unloading mechanisms, is currently under way.

2.2.3   Human Studies (Bed Rest)

In bed rest studies, the human body is positioned recumbent in special clinic beds usually with six-degree head-down tilt, HDT (Fig. 2.13), to study the effects of immobilization on the unloaded healthy human body during or after short-term- (>7 days), medium-term- (>21 days), and long-term-duration (>60, 90, or 120 days) bed rest in controlled laboratory environments such as provided by transitional bed rest wards (infirmary rooms) provided by university hospitals (Charité Berlin) and more specialized bed rest facilities such as provided by the National Aerospace Institute, Cologne (DLR, :envihab, Germany), and the Medical Space Clinics (MEDES) in Toulouse, France, in Europe. The HDT supine position results in partly unloading of the body with a concomitant body fluid shifting from lower to upper body parts (centralization of venous blood). An increased preload (venous return flow to the right atrium of the heart) in the cardiovascular system results in endocrine stimulation (atrial natriuretic protein secreted from heart atrium) with an enhanced renal fluid and mineral (Na^{2+}, Ca^{2+}) excretion that is even more intensified by bone mineral turnover and loss during the bed rest period (LeBlanc et al. 2007).

Immobilization-induced human body unloading results in reduced muscle tone and neuromuscular activity (hypokinesia) of trunk, back, and leg postural (anti-gravity) muscles which also result in typical structural changes (e.g., muscle volume and myofiber size reduction, myofiber slow-to-fast-type transition, altered

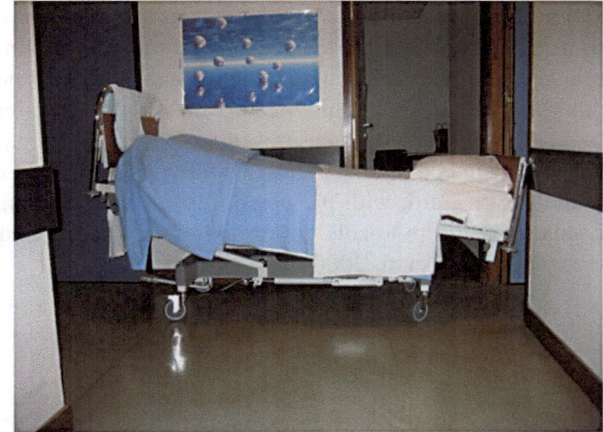

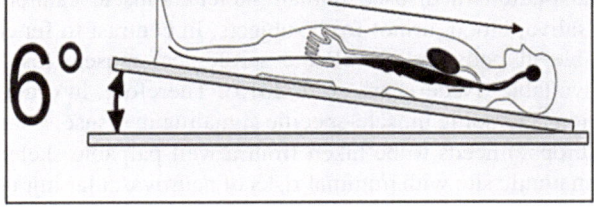

Fig. 2.13 A typical 6° head-down tilt (HDT) bed rest scenario/spaceflight analogue. *Upper panel* shows an empty bed at HDT position at the corridor of MEDES Space Clinics, Toulouse, France. *Lower panel* shows that volunteers at HDT supine body position are challenged by cephalad fluid shift (toward head) and partial unloading with disuse atrophy particularly of leg skeletal muscles throughout the entire bed rest period of weeks or months (spaceflight analogue model) (Images: D. Blottner)

capillary-to-myofiber ratios, intermuscular adipose tissue) and functional changes (e.g., fatigue resistance, loss in power and force, reduced neuromuscular activity) observed after extended muscle disuse (Clark 2009). As a consequence, body immobilization with hypokinesia in HDT results in disuse-induced loss of bone and muscle mass (estimated atrophy rate 1–3 % per week HDT) with reduced muscle strength (power and force) and, finally, also impaired performance control similar to the known challenges observed in crew members in microgravity in Space or after return to Earth. Since decades, HDT bed rest has been considered as an experimental analogue to spaceflight on the ground that is controlled more easily and with statistically sufficient high numbers of voluntary subjects at more economical costs and with less health risks than for human experiments with crew members under real microgravity environments (Convertino et al. 1989; LeBlanc et al. 1997). At end of a bed rest study, the volunteers are slowly accustomed to an upright body position with the help of a tilt table (orthostatic stimulation) and become instantly part of a post-recovery and rehabilitation protocol that usually is completed after 6 months or 1 year following bed rest and that again may allow for recovery studies after reloading of the human body comparable to gravitational reloading studies in spaceflown crew members after their return to Earth. Even though musculoskeletal and neuromuscular changes found in HDT are largely the same found in spaceflight or after return thereafter, it is quite reasonable that a number of other microgravity challenges for the deconditioned human body in real space missions (e.g., radiation, stress, proprioception, autonomic dysregulation, immunosuppression, vestibular control) or, for example, extreme physical and/or cognitive inflight challenges of crew members (mission duties, extravehicular activities) may not be adequately simulated by the bed rest paradigm with "terrestrial astronauts" at 1 G, for example, simulating μG-induced changes in postural reflexes, sensorimotor behavior, and visual-vestibular neural stimulation (Reschke et al 2009). As animal study results may not be transferred 1:1 to human physiology, the use of countermeasures of various modes (e.g., exercise, nutrition) can be tested only in bed rest for their feasibility and effects on disuse-induced muscle and bone loss on Earth with the possibility for their implementation to future inflight countermeasure protocols during long-term space missions to the Moon or Mars (Pavy-Le Traon et al. 2006).

2.2.3.1 Methods I: Muscle Biopsy

Many of the structural changes and in particular the muscle-specific cell signaling adaptation in disused human skeletal muscle cannot be studied in body fluids (saliva, blood, urine) from subjects. In contrast to functional bone markers detectable in body fluids, reliable serological muscle loss markers are presently not available (Nedergaard et al. 2013). Therefore, in order to study structural adaptations including muscle-specific signaling in disuse, a small amount of muscle tissue (biopsy) needs to be taken from a well-palpable skeletal muscle at a well-known anatomic site with minimal risks of neurovascular injury (blood vessels and nerves) and with minimal to moderate risks for discomforts (e.g., pinching pain, local hematoma, infection, scar formation). A muscle biopsy can be harvested by an experienced operator from well-palpable leg muscles of voluntary human subjects or crew members using the well-established needle biopsy technique (Bergstrom 1975). The so-called Bergström needle consists of a small hollow needle (12–15 gauge size) that is used to percutaneously harvest a small muscle tissue sample after local skin cleaning and anesthesia without or with suction modification (Tarnopolsky et al. 2011). The local anesthesia using lidocaine with epinephrine (usually injected to skin, subcutis, and fascia to help control incision-related bleeding) may, however, confound molecular analysis in needle biopsies that should be avoided by a more careful injection approach (Trappe et al. 2013). The Duchenne-Bergström percutaneous needle biopsy technique routinely gains a rice corn-sized tissue sample of about 100 mg > 150 mg of tissue (single pinch) routinely used in many previous human physiology studies, in bed rest, but also in crew members (Fig. 2.14). Currently, thinner needles are available recovering about 4 mg of tissue (single recovery) from the human vastus lateralis or from the latissimus dorsi back muscle that may open up new anatomical sites and functional muscle types to be analyzed, for example, by molecular tools where a few mg of sample amount is sufficient enough for molecular analysis (Paoli et al. 2010). Alternatively, a Rongeur surgical forceps developed to cut away bone and tough tissues has been used for a biopsy of soft muscle tissue (<150 mg) through an approx. 1 cm skin/fascia incision following local anesthesia in bed rest studies (Rittweger J, personal communication). In all cases, routine medical wound care (medical plaster) and, if necessary, anti-scar bandage with pressure dressing are applied to the small skin wound on the day of biopsy. The quality of the small tissue sample (fiber orientation, fascia and connective tissue impurities, blood coagulate)

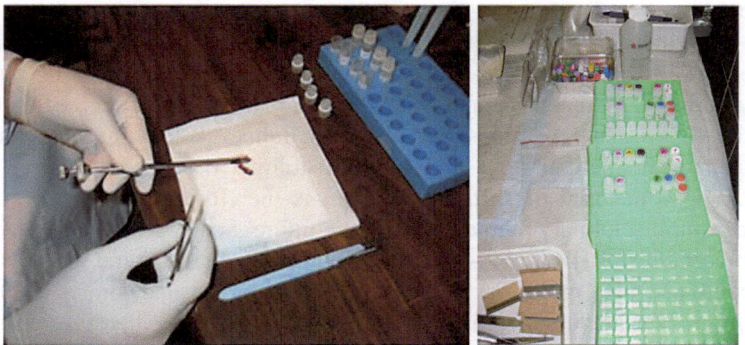

Fig. 2.14 Typical scenarios of human skeletal muscle biopsy tissue sampling (*right*, needle biopsy) and sample collection (*left*) using color-coded microcups for muscle team principal investigators (PIs). Image (*left*) taken from the 90 days LTBR study at MEDES, Toulouse, France, and (*right*) from the 60 days BBR-2 study at the Charité, Campus Benjamin Franklin, Berlin, Germany, 2007. In the BBR-2 study, an open incision biopsy procedure was used (Belavý et al. 2010) (Images: D. Blottner)

is instantly inspected under a binocular microscope, while the sample is cleaned of non-muscle tissue, cut to smaller aliquots as required (e.g., according to histology, biochemical and molecular tissue preparations), immediately frozen in liquid nitrogen (or treated for other preservation methods required), and usually stored at minus 80 °C freezer (or in 4 °C refrigerator) for further sample analysis in a laboratory on site or at home.

Due to sharp edges and scoop-shaped tips, both Bergström needle and Rongeur forceps share some risks of partial mechanical stress to the soft muscle sample that may interfere with a more broad tissue analysis including routine histology, ultrastructural analysis, and routine biochemical or sophisticated proteomic analysis. Some difficulties in biochemical and molecular measures caused by the needle biopsy protocol were reported that may result in variable changes in immunoblot signals, for example, when muscle samples were to be analyzed for cell signaling in human tissue (Caron et al. 2011). In order to minimize any mechanical stress during biopsy sampling and to improve tissue quality and quantity, an open incision biopsy technique has been alternatively used in two ESA long-term bed rest studies at the Charité Berlin (2003 Berlin Bed Rest (BBR)-1 and 2006–2007 BBR-2 study) that resulted in the sampling of muscle tissue of excellent quality (longitudinal fiber orientation) and quantity (>250 mg/biopsy) from the thigh (quadriceps femoris vastus lateralis) and the deep calf muscle (soleus). After local skin/fascia anesthesia, an approximately 1.5 cm skin/fascia incision is made, and a small bundle of muscle fibers is ligated and carefully separated from the adjacent fiber bundles of the muscle under visual control. Following proximal and distal cutting, the small ligated fiber bundle (approx. 300–400 fibers in longitudinal orientation) is then directly excised from the periphery of muscle by a surgeon, immediately handed over to another operator for further sample preparation, freezing, and storage. Meanwhile, both fascia and skin are sutured in situ (2–3 sutures) followed by normal wound care as shown in the BBR-2 study protocol paper (Belavý et al. 2010).

The open incision biopsy is a safe alternative to the standard needle biopsy but requires a well-trained surgeon and a surgery room of a clinic in a hospital. As already outlined, the latter technique has some advantages over the needle biopsy by getting "gold standard" biopsy samples from human skeletal muscle with much less mechanical manipulations and better histologic tissue preservation. High-quality biopsy material of sufficient amount is therefore proposed for comprehensive multidisciplinary human studies in future Space Life Sciences where ultrastructural/histological/high-resolution confocal laser structural analysis needs to be combined with molecular cell biology and sophisticated -omics technology (prote-/transcript-/metabol-/signal-omics) to get more comprehensive insights by using interdisciplinary tissue analysis to further understand the normal and abnormal mechanistic control and maintenance levels of disused skeletal muscle (Moriggi et al. 2010; Salanova et al. 2014) as a valuable on ground data base for comparison with the microgravity-induced changes in skeletal muscle atrophy.

The human skeletal muscle system is comprised of different muscle types with different functions (postural control/stabilization vs. fast mobilization) and different myofiber distribution patterns (type I, IIa, IIx) typical to a given muscle that appear to also respond in a typically muscle-specific way to disuse and exercise. Due to different physical training, routine daily activity patterns, endocrine status (sex hormones), and age, human skeletal muscle may show a high intersubject variability that should be considered for the study design in Human Space Life Sciences, for example, with study inclusion of sex- and age-matched subjects. For standardization, a biopsy is usually taken from at least two muscle types with a different fiber-type composition, e.g., the mixed fast/slow fiber composition thigh muscle (vastus lateralis) and the more slow fiber composition calf muscle (soleus), as reference muscles due to the fact that a wealth of structural and functional data was accumulated from these muscle types in many human studies on the ground and from spaceflight experiments over the last two or three decades.

In bed rest, a muscle biopsy is usually taken from volunteer subjects (male or female) after signed informed consent briefing before start of bed rest (pre-biopsy) and at the end or after bed rest (post-biopsy) in order to investigate short-, medium-, and long-term skeletal muscle changes and their possible prevention by countermeasures at various structural and molecular levels, for example, in bed rest subjects or crew members. If exercise interventions were applied as countermeasures in disuse on the ground or during the crew member's preflight training, the test constraints include no strenuous exercise during 24 h prior to the biopsy, no exercise on the biopsy day, or working out of the biopsied muscle/leg for at least 24 h after the biopsy. In human spaceflight experiments, usually a preflight (several weeks before launch) and one or two postflight biopsies (shortly after landing/after 1–2 weeks of recovery) might be taken from the crew members (presently using standard needle biopsy).

Inflight muscle biopsies may not be feasible in the near future in fact due to ethical and safety reasons because wound healing and the associated pathophysiological processes under microgravity conditions are not yet fully understood (Davidson et al. 1999), and thus more serious risks (inflight emergency) may not be excluded for the crew during their mission duties in Space.

2.2.3.2   Methods II: High-Resolution Confocal Laser Scanning (CLS) Microscopy

Skeletal muscle atrophy is paralleled by histological changes in myofiber size (myofiber cross-sectional area) and myofiber phenotype distribution (slow type 1, fast type 2a, 2x) that have been investigated in biopsy tissue sections (paraffin or cryosections) by standard histochemistry and immunohistochemistry methods with conventional light and fluorescence microscopy. With the development of more stable (less bleaching/fading) fluorescence signal-conjugated secondary antibodies (e.g., Alexa 400, 488, 550, 630 nm, molecular probes) in combination with double and triple immunostaining protocols in one and the same muscle cryosection, high-resolution three-channel confocal laser scanning (CLS) microscopy became the method of choice for specific antibody detection following

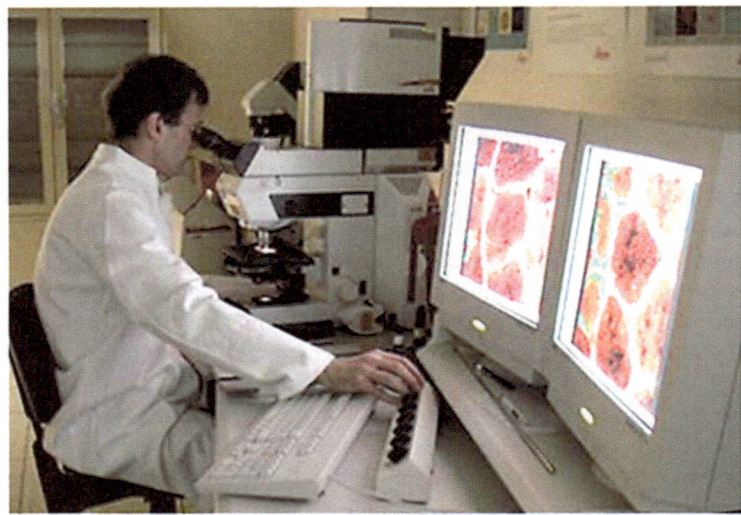

Fig. 2.15 Confocal laser scanning (CLS) microscopy device (Leica TCS SP-2, leica-microsystems.com) used in the Neuromuscular Group laboratory at the Charité Universitätsmedizin Berlin, Germany, since 2001 (DLR grant # 50WB0145 to D.B.)

multiple immunostaining protocols for high resolution of particular functional muscle cell structural and functional biomarkers relevant for various subcellular compartments (e.g., sarcolemma, sarcosol, contractile apparatus, neuromuscular junction) as well as for muscle-specific signaling pathways (Fig. 2.15). Briefly, confocal laser image analysis is achieved by the stepwise optical dissection planes through 8–30 micron (μm) thick cryosections (comparable to a high-resolution cell CT using a stable laser light source instead of X-ray). The optical dissection steps by the highly sensitive laser beam-splitter technology can be preset to variable consecutive optical planes (e.g., 0.01 μm/step), thus leading to a precise signal detection of various fluorescent biomarkers (red-green-blue) at subcellular compartments (without diffuse and nonspecific signal noise often detected with immuno-histological sections by routine epifluorescence microscopy).

2.2.3.3   Muscle and Fiber Size (CSA) and Fiber-Type Distribution (Slow/Fast) in Bed Rest

The loss in muscle mass during immobilization also termed disuse atrophy (muscle wasting) has been determined in bed rest noninvasively mainly in human postural muscles by routine clinical imaging techniques. For example, the disuse-induced structural size of whole skeletal muscle cross-sectional area (MCSA) was routinely monitored by magnetic resonance imaging (MRI) that, for instance, gives reproducible muscle volume image data sets from individual deep back muscles (e.g., multifidus) and the calf soleus known for their strong atrophy responses following longer periods of disuse (Belavý et al. 2011). In addition, whole body MRI screening in bed rest revealed that the magnitude of MCSA reduction in postural but also in other limb and trunk muscles may be region-specific and not uniform within a given muscle but may rather show variable sites of atrophy throughout the full length of a muscle (Miokovic et al 2012). Alternatively, high-resolution ultrasound (ultrasonography, echography) that may be more easily applied in bed rest has been used more recently to noninvasively monitor MCSA in normal subjects (Esformes et al. 2002), following high-intensity muscle training (Ahtiainen et al. 2010), and in bed rest (Arbeille et al. 2009) with reproducible and valid results and comparable degree of precision to MRI.

Nevertheless, in vivo anatomical variability may be found in axial cross sections, for example, at mid-thigh or calf bone, muscle, and related neurovascular structures (regional nerves and blood vessels) at the human limbs often showing region-specific interfascial tissue spaces or variable intermuscular and intramyocellular fat infiltrations or deposits. It is also known that acute or strenuous exercise in normal subjects usually results in post-training edema that may last for up to several hours following exercise intervention, for example, reported for the human thigh muscle (Ploutz-Snyder et al. 1997) which may also likely occur in long-term bed rest following strenuous exercise interventions as countermeasures. In addition, significant differences in size measurements were reported between recumbent versus standing MRI measures (Berg et al. 1993), and the known cephalad (head-directed) fluid shift from the legs to the upper body during supine bed rest (Thornton et al. 1992) may still provide a yet underestimated bias in the morpho-metric analysis of whole skeletal muscle size, for example, by X-ray-based technology (Rittweger et al. 2013). Unlike for bones, tendons, or even strong fascia layers, the current available routine and more advanced digital imaging technology for soft tissues based on gray-scale determination (MRI, ultrasound, spectroscopy, X-ray based) still does not provide adequate signal-to-noise resolution criteria for the clear signal discrimination between the various body soft tissue compositions and their individual components such as water, fat, blood, or lymph fluid versus connective and soft tissue that may at least partly add to the variable size changes particularly found in disused human limb tissue including skeletal muscle and support tissue structures in long-term bed rest. Yet limited evidence for reliability of the current MRI measures or even of real-time ultrasound measures in routine human limb muscles size monitoring in clinical settings (English et al. 2012) as well as in bed rest immobilization is still critical. There is a need for advanced technology with, for example, a better signal resolution and discrimination between more discrete signals of various biological tissue components to properly study atrophy and size changes in addition to noninvasive monitoring of muscle volume changes that are required in future bed rest studies. Though the presently available MRI techniques used in routine clinical imaging may be helpful for diagnostic purposes, they may only show larger volume changes of the whole muscle tissue (and its non-muscle components and fluid content) under prolonged bed rest conditions without excluding the obvious bias of over-/underestimations of the actual magnitude of structural myofiber atrophy actually found in the various fibers

of intramuscular fascicles and, more importantly, detectable by variable size
changes of individual myofiber phenotypes (slow, fast), for example, only by
high-resolution confocal laser scanning microscopy.

2.2.3.4 NO/NOS as Biomarker to Study Efficacy of Physical Exercise Countermeasure

Emerging evidence accumulated showing that biological NO signals have multiple
functions in normal skeletal muscle including muscle contractility, soreness, and
fatigue that altogether are typically known as skeletal muscle functional impairments following various periods of disuse on the ground or that have been reported
from crew members during spaceflight missions and following recovery thereafter
(see Sect. 2.1.6.). We therefore hypothetized that NO signaling may be altered in
normal human skeletal muscle following disuse and that changes in NOS/NO can
be ameliorated if not prevented by physical exercise as countermeasure during the
periods of disuse in bed rest. In particular, we were interested to know if NOS/NO
could be used as a reproducible and reliable structural and functional biomarker for
muscle activity versus inactivity for testing of the efficacy of an exercise countermeasure protocol during bed rest disuse.

One of the first evidence of atrophy-induced changes in NOS1 in normal human
skeletal muscle was reported from our immunohistochemical analysis of all three
NOS isoforms (NOS1, NOS 2, NOS3) in muscle biopsy material originating from a
90-day ESA long-term bed rest study performed in 2001 at the MEDES Space
Clinics in Toulouse, France (Rudnick et al. 2004). In this study, we showed
substantial loss in NOS proteins in human muscle biopsy lysate preparations from
slow- and fast-type muscle (VL, SOL) of the bed rest control group (bed rest, no
exercise) compared to normal control (before bed rest). By contrast, maximal force
resistive exercise countermeasure performed in a bed rest resistive exercise
(RE) group (fly wheel technology, 3×20 min/weekly) prevented such changes
almost completely in both VL and SOL suggesting that the quantity of muscular
NOS is activity driven and that in principle exercise is able to maintain the NOS1
myofiber status in human skeletal muscle in longer periods of disuse (Fig. 2.16).

We also found in this study that NOS2 (inducible NOS) is co-localized to
caveolin-3 (important signal component of the muscle membrane) at the human
sarcolemma and is highly regulated by resistive exercise and NOS3 (endothelial
NOS) is upregulated by resistive exercise in the microvascular/capillary bed sur-
rounding the muscle fibers in the trained bed rest group (Rudnick et al. 2004). These
results were largely confirmed by the second ESA Berlin Bed Rest Study in 2003,
BBR-2 (Blottner et al. 2006), and also following 7 and 14 days' human unloading
study of "dry immersion" (Koryak 2002) without or with plantar support stimulation in a cooperation experiment between the Charité Berlin and the IMBP, Moscow
(Moukhina et al. 2004). Briefly, plantar stimulation appears to be a very effective
way of dynamic biomechanical stimulation applied to the skin mechanosensors
(PACINI bodies) of the foot sole of the unloaded human leg that positively affects

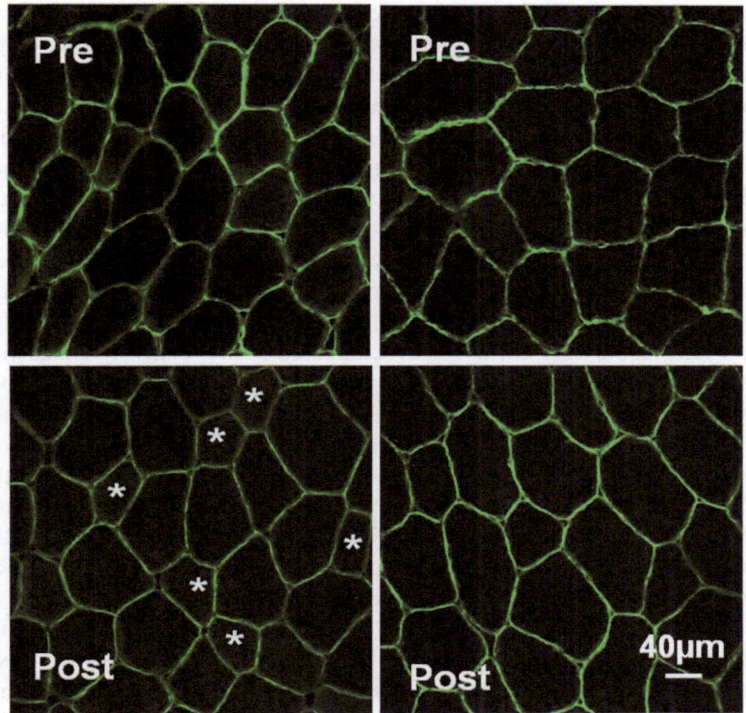

Fig. 2.16 Sarcolemmal NOS1 immunostaining in human skeletal muscle fibers before (Pre) and after bed rest (Post) without (*left column*) and with exercise (RVE) countermeasure (*right column*). The intensity of sarcolemmal NOS1 immunosignals is maintained after exercise (Rudnick et al. 2004). Asterisks denote atrophic myofibers with diffuse cytosolic NOS1 immunostaining (60-d BBR-1 study, Charité Berlin, Germany)

the calf muscle tone and contractile properties while the body is floating "dry" in a water tub (dry immersion) separated by a thin water-resistant membrane established and performed at the Institute of Biomedical Problems (IMBP) of the Russian Academy of Sciences in Moscow, Russia (Koryak 2002).

In animal studies, muscle-specific splice variant μNOS delocalization inhibits muscle force in dystrophin-null mice suggesting nitrosative stress elicited by NO generated by NOS as inhibitor in muscle force generation (Li et al. 2011). Knockout (KO) experiments ($\mu NOS^{-/-}$) in mice resulted in myopathy and reduction in contractile force, thus underlining the key role of NOS in muscle force production (Percival et al. 2010). Two animal studies furthermore suggested involvement of endogenous NO production inhibited by L-NAME (potent inhibitor of NOS) in chronic overload-induced skeletal muscle hypertrophy of rat plantaris muscle and myofiber-type transition (Smith et al. 2002) or in activity-induced calcineurin-NFATc1 signaling (involved in activity-dependent myosin heavy chain isoform expression) and fast-to-slow skeletal fiber-type conversions (Martins et al. 2012).

In normal human skeletal muscle, we also observed an altered expression of sarcolemmal NOS following longer periods of disuse and exercise. After long-term bed rest (LTBR 2001), for example, we found that the loss of sarcolemmal NOS localization in atrophied human soleus myofibers can be prevented by maximally resistive exercise as countermeasure (fly wheel technology) compared to a non-exercising bed rest control group suggesting that sarcolemmal NOS is activity dependently expressed in human skeletal muscle (Rudnick et al. 2004). In a 60-day bed rest study (1st Berlin Bed Rest Study, BBR-1, 2003, Blottner et al. 2006), we showed that the sarcolemmal NOS expression can be maintained by whole body resistive vibration (RVE) which represents a short but high-intensity stimulation training protocol applied to subjects during bed rest (approx 10.000 muscle contraction cycles 2–3x/week for 3–5 min) compared to the relatively exhaustive RE fly wheel technology (performed for 20 min for 3 days/week). Thus, vibration may be a highly effective biomechanical signal for human muscle that may be useful to maintain functional NOS following longer periods of immobilization such as in bed rest or during real spaceflight. In a follow-up 60-day bed rest study (2nd Berlin Bed Rest, 2006), we compared the training effects in bed rest between two different resistive exercise RE groups, RE only (RE without vibration) and RVE (RE plus vibration), on various structural skeletal muscle parameters. Notably, both RE and RVE protocols performed on the same Galileo Space® trainer (MediTec, Pforzheim, Germany) maintained sarcolemmal NOS localization patterns supporting the general idea of the presence of an activity-driven NOS expression. If similar findings were obtained from crew members in real spaceflight, it will be investigated in future planned work.

Disruption of the subsarcolemmal localization of NOS1 (nNOS) appears to be a signature of fiber atrophy in animal and human normal skeletal muscle fibers with as yet unknown consequences (Miyagoe-Suzuki and Takeda 2001). A more recent report on the NOS sarcolemma-sarcosolic translocation mechanisms suggested that the cell permeable, saline-manganese compound EUK-134 (Lawler et al. 2014) ameliorates NOS translocation in 54 h hind-limb suspended adult F344 rats which also suggest involvement of redox signaling that may be either decreased or increased following muscle unloading (2.2.3.5, nitrosative stress). From fundamental animal studies on ground as well as in microgravity, we suggest that the quantity changes of sarcolemmal NOS expression at the outer muscle fiber membrane likely reflected some fundamental mechanistic principles and their changes in the molecular microdomain membrane composition in skeletal muscle disuse that need to be further investigated. A more recent work shows that the stress protein/chaperone GRP94 which binds to NOS mechanistically stabilizes the NOS multiprotein complex at the myofiber sarcolemma (Vitadello et al. 2014). The changes in sarcolemmal NOS expression have been also reported in various human skeletal muscle dystrophies (e.g., Chao et al. 1996; Crosbie et al. 2002; Tidball and Wehling-Henriks 2004; Fanin et al. 2009), suggesting aberrant sarcolemmal NOS localization patterns as valuable pathophysiological markers in various myopathies that may be helpful in their clinical diagnosis (Zhou and Zhu 2009).

In spaceflown mice skeletal muscle (see Sects. 2.2.2.1 and 2.2.2.2), we found that translocation of sarcolemmal NOS to sarcosol is particularly apparent in atrophic soleus but not in the extensor digitorum longus (EDL) compared to ground controls (Sandona et al. 2010) supporting coexisting mechanisms of disuse atrophy and NOS translocation are signatures to microgravity disuse atrophy.

2.2.3.5   Skeletal Muscle Say "NO" to Disuse on Earth: NO and Nitrosative Stress in Human Skeletal Muscle

Nitric oxide synthase (NOS) and the diffusible NO signals are expressed in animal and human skeletal muscle with pleiotropic biological roles for normal skeletal muscle structure and function (Bredt and Snyder 1994; Kobzik et al. 1994). Depending on muscle activity or inactivity, muscular NOS proteins and NO signals are either up- or downregulated; NOS1 is translocated in myofibers from sarcolemma to the sarcosol in animal and human muscle disuse and myopathies, as well as in spaceflown mice (see Sects. 2.2.2.1 and 2.2.2.2). Uncontrolled over-/underproduction and/or aberrant expression of free radicals (Lawler et al. 2003) such as superoxide (O_2^-) or nitric oxide (NO) in skeletal muscle tissue provide the basis for oxidative/nitrosative stress that is induced by reactive oxygen/nitrogen species (ROS/RNS).

Nitrosylation of proteins is characterized by the covalently binding of NO to reactive free cysteine residues ($-SH$ groups) of a protein that are instantly converted to S-NO groups recently identified and listed by functional target SNO-proteins of the SNO-proteome (Seth and Stamler 2011). In analogy to phosphorylation and acetylation of functional cell peptides/proteins, the term nitrosation or S-nitrosylation has been proposed for SNO-protein modifications (Fig. 2.17) which are supposed to

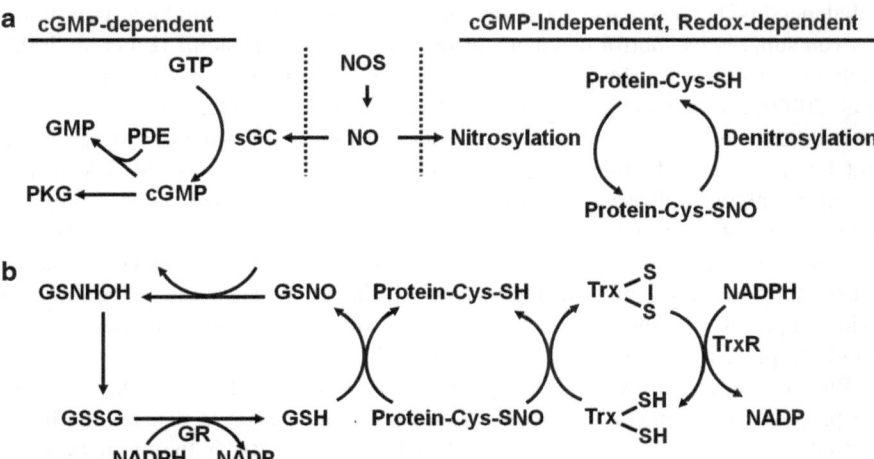

Fig. 2.17 Scheme showing protein-S-nitrosylation (SNO-proteins) via NO-based signaling. Nitrosylation of proteins (protein-Cys-SNO) is a cGMP-independent but redox-dependent mechanism that is (**a**) driven directly by NO (*upper middle panel*) and (**b**) by glutathione (GSH, *left panel*) and thioredoxin reductase (TrxR, *right panel*), for reversible protein nitrosylation/denitrosylation (*lower middle panel*) via NADP-dependent redox mechanisms (Lima et al. 2010)

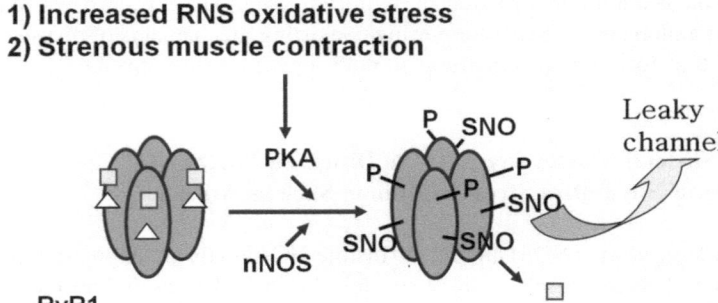

1) Increased RNS oxidative stress
2) Strenous muscle contraction

Fig. 2.18 Proposed model of nitrosative stress by reactive nitrogen species (RNS) for decreased exercise capacity. Ryanodine receptor proteins (RyR1) are hyper-nitrosylated (SNO) via nNOS and hyperphosphorylated via protein kinase A (PKA) resulting in leaky channels with myocellular calcium imbalance

have a tissue half-life of approximately a few seconds to a few minutes maximum in biological tissues (Foster et al. 2009). Under normal physiological conditions, S-nitrosylation is a reversible posttranslational qualitative modification of functional proteins and a major source of NO bioactivity, thus regulating cell dynamics and plasticity in a variety of tissues and cells (Hess and Stamler 2011). In many cell types and tissues including skeletal muscle, also the quantitative nitrosylation via unbalanced NO cell signals (hyper-/hyponitrosylation) either stimulates, maintains, or even suppresses control mechanisms of functional cell proteins (Eu et al. 2003) such as those shown for the myosins of the contractile apparatus (Nogueira et al. 2009), for the sarcoplasmatic reticulum (SR) calcium release channel protein ryanodine receptor type-1 (Fig. 2.18, Bellinger et al. 2008; Salanova et al. 2008), and for a number of other calcium release channels (Aracena et al. 2005). For example, calcium release channel proteins such as RyR1 may become "leaky" in the presence of too much of NO which may result in muscle stiffness and lack of force production (Powers et al. 2007). Muscle disuse atrophy also induces oxidative stress that accelerates protein degradation via increased substrate recognition of oxidized proteins by the E3 ligases (MuRF1-3) that, in turn, promote their susceptibility to proteolysis (Powers and Lennon 1999; Du et al. 2004).

    We were able to show that activity-induced NO imbalance and nitrosative stress are proposed mechanisms of disrupted calcium homeostasis in disuse atrophy of normal human skeletal muscle in bed rest (Salanova et al. 2013). In muscle biopsies obtained from the 60-day BBR-2 study at the Charité Berlin, Germany (2007–2008), we investigated differential levels and expression of functional SNO muscle proteins of the excitation-contraction (EC)-coupling machinery related to the control of calcium balance in muscle, for example, ryanodine receptor type 1 (Salanova et al. 2008, 2009), but also voltage-dependent calcium release (DHPR1a) and calcium uptake proteins (SERCA1&2, PMCA1) in chronically disused normal skeletal muscle in bed rest using a modified biochemical assay for the detection of SNO-proteins in cell and tissue samples (biotin-switch technique, Jaffrey and Snyder 2001) in combination with high-resolution confocal image analysis and real-time quantitative polymerase chain-reactivity (qPCR) analysis of Nrf-2 transcripts, a master gene which activates the transcriptional activity of antioxidant responsive elements (AREs) of many cytoprotective genes in the cell nuclei, in soleus, and in vastus lateralis muscle fibers in bed rest following two types of resistive exercise, RE and RVE, as countermeasure (Salanova et al. 2013). Compared to control baseline levels (bed rest start) and to the non-trained bed rest control group (bed rest end), most of the functional SNO-proteins investigated in the exercise group (RyR1, SERCA, PMCA, myosins) were downregulated, while others were maintained following either of the exercise protocol after chronic disuse in bed rest. Interestingly, variably nitrosylated NOS1 (SNO-NOS1) protein levels were found after chronic disuse in bed rest (SNO-NOS1 decreased) and following exercise countermeasure during bed rest (SNO-NOS1 increased) suggesting also NOS1 being a novel candidate in the growing list of muscular SNO-proteins (Fig. 2.19) that share activity-driven autonitrosylation mechanisms to occur with as yet unknown functional significance for the physiological and molecular control of a NOS1 activity and myocellular NO production in a muscle-specific way (Salanova et al. 2013).

Like other SNO-proteins, we propose that NOS1 may be controlled in skeletal muscle fibers, for example, by S-nitrosylases and denitrosylases (Jaffrey et al. 2001; Seth and Stamler 2011). Previous animal studies reported protein nitrosylation to be

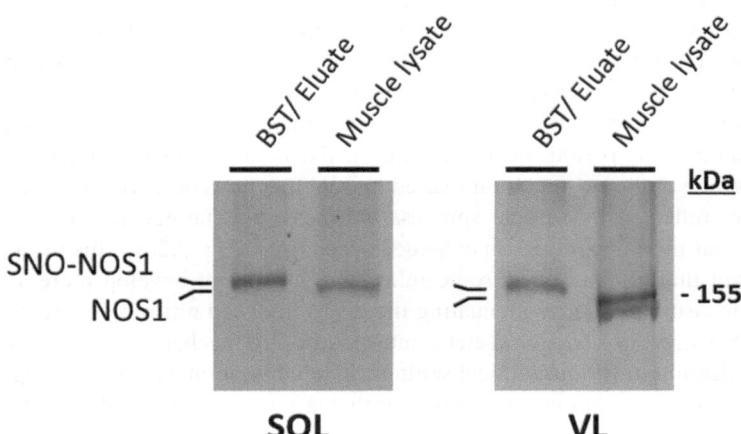

Fig. 2.19 The NOS1 protein itself is S-nitrosylated as shown by molecular weight shift of SNO-NOS1 seen in human soleus (*SOL*) and vastus lateralis (*VL*) protein BST/eluates compared to controls (muscle lysate) identified by a modified biochemical BST-based protein-S-nitrosylation assay (Salanova et al. 2013)

responsible for increased in vivo proteolysis in cirrhotic rats (Wang et al. 2010). Nevertheless, aberrant levels of functional SNO-proteins are a signature of differential nitrosative stress management in atrophic/disused versus trained human skeletal muscle related to calcium balance (Salanova et al. 2013).

As already shown for reactive oxygen species in exercise biology (Ji 2008), a contraction-induced overproduction of reactive nitrosative species and nitrosative stress in muscle likely occurs following maximally heavy loading bouts under extreme training conditions (overload) or by extremely long-duration exercise (marathon runners) in sports which might explain some of the changes observed in remarkably small subsets of muscle genes (P13K/Akt/mTOR, Foxo transcription factor, atrogin-1, MuRF1/Mafbx, myostatin) that, when turned on (activated) by unloading/disuse, may trigger discrete molecular pathways that lead to atrophy in various atrophy models (Urso et al. 2009) or that, when turned off (suppressed) by normal muscle activity and by exercise, may trigger molecular signaling mechanisms of structural muscle fiber preservation, such as size and phenotype distribution, and functional muscle fiber parameters, such as contractility, reduce fatigue, and/or maintain force production in normal and exercise physiology (Suhr et al. 2013). A current report addressing the NOS/NO signaling pathway mechanisms suggested that the NOS1-activated NADPH oxidase NOX4 may contribute to fiber hypertrophy in the overloaded skeletal muscle (Ito et al. 2013). Likewise, uncontrolled excess of reactive nitrogen species and nitrosylated functional proteins (SNO-proteins) including auto-nitrosylated NOS1 (nNOS), as, for example, reported from our laboratory after bed rest immobilization may equally result in adverse nitrosative stress conditions, thus triggering disuse-induced muscle atrophy mechanisms (Salanova et al. 2013). These findings well correlated with the results reported by Bellinger et al. (2009) showing that dystrophic muscle is characterized by the presence of hyper-nitrosylated RyR1 leaky channel and by Lehnart et al. (2005) who showed that specific deletion of a cAMP-dependent phoshodiesterase (PDE4D3) in a mouse model results in expression of PKA-dependent RyR2 hyper-phosphorylated leaky channels.

In fact, unbalanced nitrosative stress management (Stein 2002; Stein and Leskiw 2000) in inactivity periods of skeletal muscle due to unloading/disuse on Earth may likely be one of the reasons for impaired muscle functions observed in Space as, for example, reflected by muscle soreness, stiffness, and fatigue in crew members during their missions' duties in extended spaceflight (Fig. 2.20). This most important novel finding may be of some relevance in order to develop more adequate countermeasure protocols attenuating the disuse-induced nitrosative stress mechanisms that may also trigger skeletal muscle atrophy mechanisms following body immobilization in various clinical settings, in rehabilitation, and also during microgravity exposure in human spaceflight or thereafter (Salanova et al. 2013).

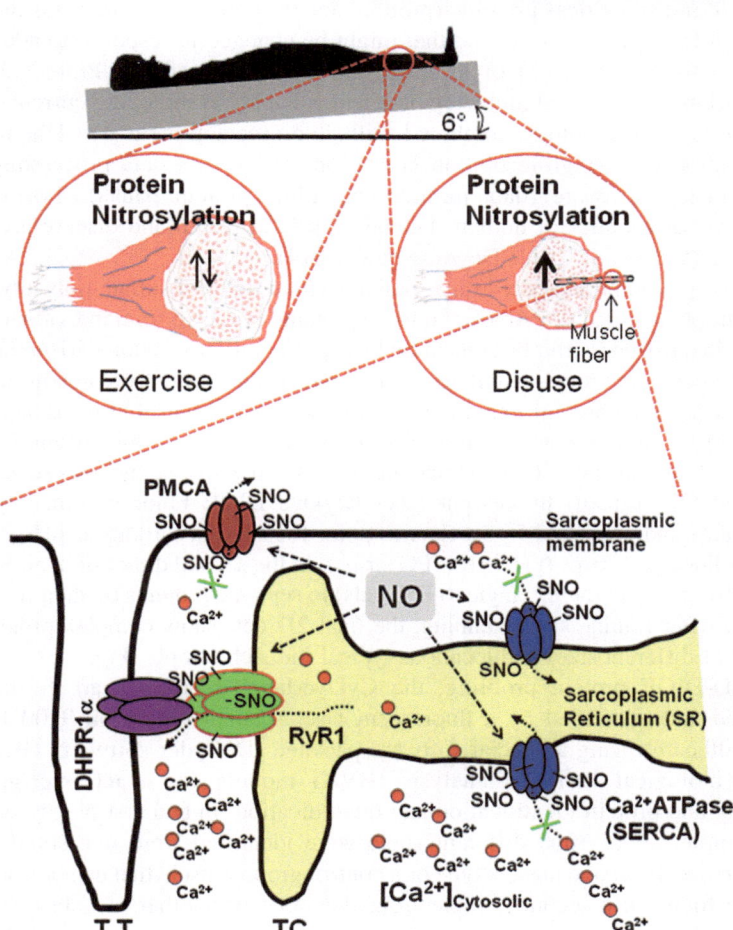

Fig. 2.20 Cartoon showing nitrosative stress management in disused human skeletal muscle studied by bed rest immobilization. Protein nitrosylation (SNO) was found to be upregulated in disused skeletal muscle and maintained at baseline following exercise countermeasure in bed rest. Lower panel shows some functional SNO-proteins (each functional receptor/channel is reflected by four subunits shown by different colors) identified in skeletal myofibers (PMCA, DHPR, RYR1, SERCA) in normal and disused human skeletal muscle (Salanova et al. 2013)

2.2.3.6 Skeletal Muscle Protein Mapping by Proteomics in Bed Rest

Proteomics is a very powerful *-omics* tool for informative mapping of the skeletal muscle protein machinery which comprises, for example, structural, contractile, metabolic, stress response, signaling, transport proteins, and probably many others.

Skeletal muscle is normally characterized by expression patterns of different structural and functional proteins (protein signature of a given muscle) that may be typical for a given muscle and that might be changed in response to adaptation processes following disuse or exercise but also in spaceflight. Proteomics is a complex laboratory-based multistep biochemical analysis including current robotic and automated technology combined with fluorescent protein dye labeling and identification using high-resolution laser scan readers and data processing using bioinformatics software (data mining via online protein banks). Two current reviews on proteomics of human skeletal muscle in health and disease are found elsewhere (Gelfi et al. 2011; Dowling et al. 2014).

Briefly, proteomics is based on a routine sodium dodecyl sulfate-polyacrylamide gel electrophoresis (SDS-PAGE) of muscle proteins extracted from individual biopsy samples in a frozen mortar by sonication in a special buffer solution (Tris/urea buffer with detergent and protease inhibitor) in two dimensions (2D). The proteins are then separated from nonprotein impurities and resuspended in lysis buffer (protein solubilization) for protein concentration determination. In the first dimension, identical amounts of proteins (40–50 µg extract) are pre-stained with cyanine dyes such as Cy3 (red) and Cy5 (green) fluorescent label (CyDye DIGE Fluor minimal dye, GE Healthcare) and separated by isoelectric point focusing according to pH gradients with nonlinear gel strips (pH 3–10, IPG strips). In the second dimension, an SDS gel (20×20 cm) is run (at 90° angle to IPG gels) to separate proteins by their molecular weights. After routine silver staining, the final 2D gels show complex protein spot patterns of different sizes and locations typical for each sample (Fig. 2.21).

In 2D-DIGE protein profiling, the CyDye-labeled proteins are furthermore identified and quantified by a fluorescent laser scanner (Typhoon 9200 Imager, GE Healthcare) using a special software package (DeCyder software, GE) and a special biological variation analysis (BVA) module for exact inter-gel spot matching and protein identification and quantification. In bed rest biopsy samples, for example, about 3000 different spots were identified from one small biopsy sample either from one muscle type or a control group muscle that can be compared to those found in a second sample from the same or another muscle type of an experimental group as, for example, required for bed rest studies (Moriggi et al. 2010). In several next consecutive steps of analysis, the area of a protein of interest is punched from the 2D gels (Ettan spot picker robotic system, GE), and the released peptides are subjected to reverse-phase chromatography (Zip-TIP C18 µ, Millipore), further processed by mass spectroscopy (MS) for protein/peptide identification, then compared with *Homo sapiens* NCBInr Database bank (verification), and finally identified by correlation with entries in SWISS-Prot TrEMBL using ProteinLynx Global Server (Waters). In combination with MALDI-ToF/MS and HPLC ESI, the 2D-DIGE method thus provides differential maps of muscle proteins and peptides from normal versus disused muscle that give new information on the quantity and quality of protein changes (several thousands of proteins) from the proteome of a skeletal muscle of interest under normal and experimental conditions (Fig. 2.21).

In cooperation with our Italian partners of the University of Milano, Italy, proteomics (2D DIGE and MS) was used in combination with a histologic analysis

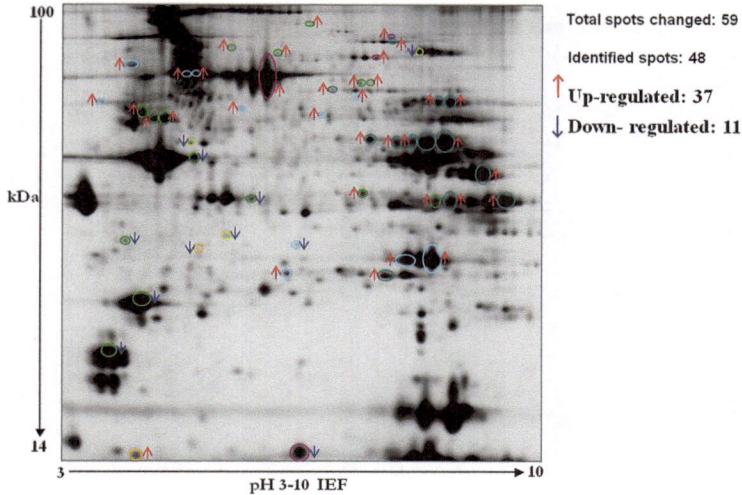

Fig. 2.21 Representative 2-DIGE protein mapping of human skeletal muscle proteome changes from soleus muscle biopsy (BBR-1 study). In the first dimension, muscle proteins were separated (first dimension) by isoelectric focusing (IEF, pH 3–10, x-axis) followed by (second dimension) molecular weight (kDa 14–100, y-axis) protein separation. From a total of 59 changed silver-stained dark spots with 48 identified spots, 37 spots were upregulated (↑, *red arrows*) and 11 spots were downregulated (↓, *blue arrows*). Individual proteins or peptides of interest are further identified by MALDI ToF/mass spectroscopy and HPLC/ESI (Courtesy: C. Gelfi (see Moriggi et al. 2010; Salanova et al. 2014))

to investigate changes in human skeletal muscle and their muscle fibers following disuse in samples from the 60-d BBR-1 study without and with vibration exercise (RVE). After bed rest, we found a substantial downregulation of proteins of the aerobic metabolism that was reversed by RVE, more in VL than in SOL. By contrast, proteins of anaerobic glycolysis were upregulated following RVE compared to bed rest without exercise countermeasure. Structural proteins from the sarcomere and the costamere microdomains were highly dysregulated by disuse and exercise (Moriggi et al. 2010).

A more detailed analysis of the structural, ultrastructural, and proteomic analysis was performed in our group with biopsy samples (SOL, VL) from the follow-up 60-d BBR-2 study (Salanova et al. 2014). In this study, we correlated some of the changes found in the proteome profiles in disused versus trained muscle in bed rest with the structural and morphometric findings from the same biopsy material following two different exercise modalities (RE vs. RVE) versus a control bed rest-only group after 8 weeks in bed rest (Salanova et al. 2014). Noteworthy, the contractile cytoskeletal and costameric proteins (tenascin-C, ROCK-1, FAK) were normalized following RVE particularly in SOL; however, costameric proteins increased by 59 % (RVE) and by 108 % (RE) versus control (no exercise). The unexpected ultrastructural damage found in bed rest after RE, but not after RVE (Fig. 2.22), was confirmed by concomitant increase in MuRF1 proteins (proteolysis biomarker) in the same

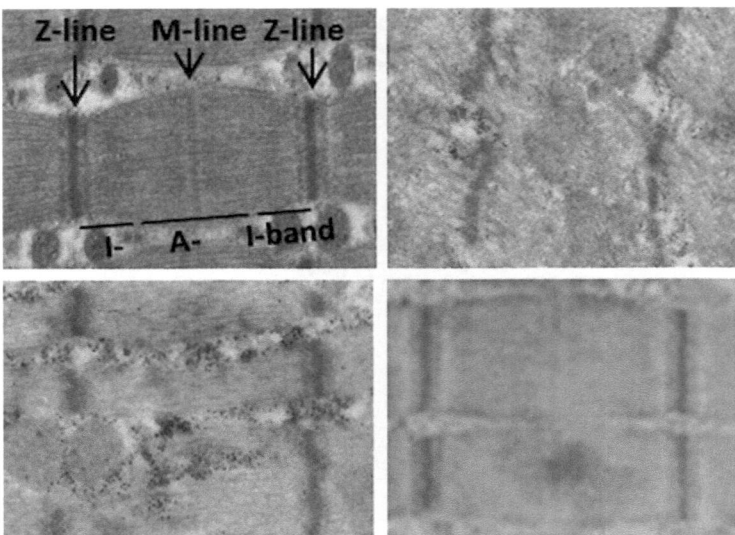

Fig. 2.22 Electron microscopy of normal sarcomeres (*upper left*) in healthy human skeletal myofibers with regular ultrastructure (regular Z-line, M-line, A-, and I-bands) and abnormal sarcomeres deteriorated after disuse atrophy in bed rest (*upper right*) showing perturbed ultrastructure. *Lower panel* shows two images of sarcomeres from soleus at identical magnifications following bed rest, with resistive exercise (RE, *lower left*) still showing signs of myofibrillar desintegration and with resistive exercise superimposed with vibration (RVE, *lower right*) with apparently regular myofibrillar orientation and normal sarcomere ultrastructure (Modified from Salanova et al. 2014)

samples. In VL, the outcome of RVE and RE on the proteomic pattern of some sarcomeric proteins (troponin, MYL2, sarcosin, desmin) was similar. As previously shown in the BBR-1 study by comparison between RVE and a bed rest control group without exercise (Blottner et al. 2006), we now provided further molecular and proteomic evidence suggesting that RE superimposed with vibration stimulation (RVE) is a highly efficient exercise countermeasure protocol against muscle atrophy, ultrastructural damage, and molecular dysregulation of basic functional muscle proteins (i.e., proteomic profiling of contractile, cytoskeletal and costameric proteins) induced by chronic disuse (Salanova et al. 2014).

In conclusion, the use of proteomics combined with structural, confocal, and biochemical analysis performed on the same biopsy material has broadened our current view on human skeletal muscle protein changes following disuse and exercise and therefore has become an inevitable and indispensable novel tool in fundamental and applied research in Human Space Life Sciences. In the near future, *omics* profiling is thought to become an indispensable new standard tool for personalized medicine in human space flight to individualize countermeasures that enhance health, safety, and performance of mission crew members (Schmidt and Goodwin 2013).

2.2.3.7   Good Vibrations Are Half the Battle Against Disuse Atrophy: Resistive Loading with Neuroreflexive Stimulation Using the Galileo® Muscle Exercise Regimen

If properly used in adequate muscle exercise protocols, vibration mechanical signals targeted to the musculoskeletal system result in exceptionally high rates of contraction cycles (approximately 1,000 cycles per second) controlled by spontaneous muscle stretch reflexes that help to mitigate disuse atrophy and reduction of neuromuscular activity in bed rest (Rittweger et al. 2010). One of the advantages of this exercise modality already addressed as RVE countermeasures in two recent bed rest studies, BBR-1 and BBR-2 (Fig. 2.22), described in previous paragraphs for normal but also disused skeletal muscle over other standard physical exercise protocols probably is the combination between low-frequency-controlled resistive loading stimuli targeted to individual muscles or groups (to maintain muscle mass) and to induce adequate neuromuscular activation (to maintain responsiveness of motor units) thereby addressing two of the well-known structural and functional negative outcomes following extended periods of disuse on Earth or in Space (Edgerton et al. 2001).

In principle, vibration signals of the RVE protocol are transmitted by an alternating tilted foot platform to individual muscles or groups of muscles of the human body either in normal standing/squatting position or even in supine body position for only short training intervals (3–5 min) on the ground (Fig. 2.23). The mechanical and neuronal signals thus triggered are equivalent to the biodynamical processing mechanisms during normal muscle activity control, for example, in human posture and gait control (Ritzmann et al 2014), that is spontaneously triggered almost automatically in a physiological way by preprogrammed spinal reflex control mechanisms. If combined with standard exercise protocols, this new physiological exercise modality has the capacity to improve the training outcome particularly for postural muscle groups compared which often require strenuous and powerful training bouts with maximally strength output with a strong compliance and level of motivation needed for users during longer training sessions for success. This may be of particular importance for the health and fitness and rehabilitation of older adults and other special populations but more and more also for the younger generation, thus facing sedentary lifestyles. Currently, frequency-controlled whole body vibration is considered to serve as an adequate physiological trigger to support neuromusculoskeletal structure and function and to stabilize muscle endurance critical to body gait and postural control (Rittweger 2010).

In view of alternative and more efficient physical exercise protocols as countermeasure to skeletal muscle disuse atrophy, resistive vibration exercise (RVE) was successfully tested to support muscle and bone quality and to mitigate a number of different disuse atrophy changes in two ESA long-term bed rest studies performed at the Charité Berlin, Germany (Rittweger et al. 2006; Blottner et al. 2006; Mulder et al. 2006; Belavý et al. 2009, 2012; Miokovic et al. 2011; Buehring et al. 2011). In principle RVE is able to address either individual muscle groups or functional

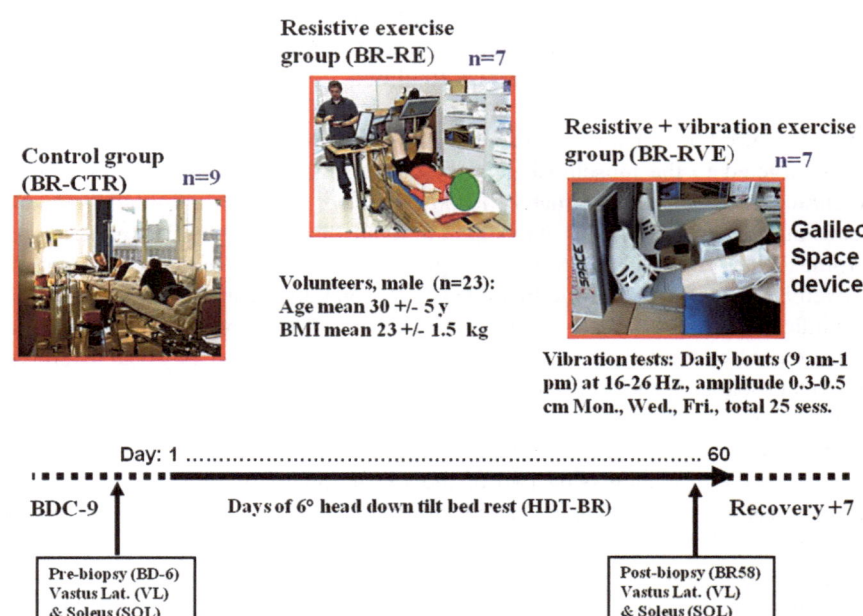

Fig. 2.23 Scenario of the 60 days BBR-2 bed rest study performed at the Charité Berlin, Germany, in 2006/2007 with three study groups, one supine control (BR-CTR) bed rest-only without exercise, one bed rest resistive exercise group (BR-RE), and one bed rest resistive exercise superimposed by vibration (BR-RVE). The muscle biopsy was taken at baseline data control (BDC-6) before bed rest and shortly before end (BR + 58) of the bed rest period. For the BBR-2 study protocols, see Belavý et al. 2010

chains of muscles (arms or legs) or to address global muscle groups of the whole human body depending on the training objectives (www.galileo-training.com). As pointed out already in the introductory paragraph, whole body vibration (RVE) is based on relatively short intervals (3–5 min) of low-frequency mechanical impulses (<20–30 Hz) with moderate amplitudes (0.5–1 cm) produced by a tilted vibrating platform for use in upright body position (Galileo®) or at supine body position (Galileo Space® trainer, Fig. 2.23), both designed and custom-made by a medico-technical company (Novotec Medical, Pforzheim, Germany). Unlike other training devices with vertical (up and down) movements, the Galileo tilted plate movements simulate human gait in a physiological way because of a side-alternating acceleration of the legs, pelvis, and lumbar spine and their individual extensor and flexor muscles (www.galileo-training.com).

The muscle contractions thus induced by frequencies starting from about 12 Hz onward are considered to be controlled by neuroreflexive mechanisms (spontaneously contractions executed via H-reflex-based afferent/efferent stimulation upon mechanical stretch signals, Kramer et al. 2013) rather than by voluntary muscle activation (voluntarily, contractions are executed by arbitrary/planned channeling from higher brain areas to efferent spinal motor activity). Because RVE makes use of preprogrammed muscle activation at the spinal cord level (spinal reflexes), the work-out efficiency (training effects) may be less influenced by the compliance/ motivation of users (e.g., users with reduced drive for motions) compared to other standard RE protocols to counteract disuse atrophy that usually require a strong mental focus on maximal power and force production during the training bouts (particularly in more or less fully body relaxation in bed rest and in spaceflight). Short RVE training intervals with, for example, 25 Hz for 3 min produce muscle stretch reflex contractions equivalent to about 4500 steps (www.galileo-training. com) which is about half of the battle with regard to the number of total steps recommended for daily physical activity in apparently healthy sedentary and older populations on Earth (Zehr and Stein 1999; Abdelmoity et al. 2000; Tudor-Locke et al. 2011).

References

Aas V, Bakke SS, Feng YZ, Kase ET, Jensen J, Bajpeyi S, Thoresen GH, Rustan AC (2013) Are cultured human myotubes far from home? Cell Tissue Res 354:671–682

Abdelmoity A, Padre RC, Burzynski KE, Stull JT, Lau KS (2000) Neuronal nitric oxide synthase localizes through multiple structural motifs to the sarcolemma in mouse myotubes. FEBS Lett 482:65–70

Ahtiainen JP, Hoffren M, Hulmi JJ, Pietikäinen M, Mero AA, Avela J, Häkkinen K (2010) Panoramic ultrasonography is a valid method to measure changes in skeletal muscle cross-sectional area. Eur J Appl Physiol 108(2):273–279

Alkner BA, Tesch PA (2004) Knee extensor and plantar flexor muscle size and function following 90 days of bed rest with or without resistance exercise. Eur J Appl Physiol 93:294–305

Anderson JE (2000) A role for nitric oxide in muscle repair: nitric oxide-mediated activation of muscle satellite cells. Mol Biol Cell 11:1859–1874

Andreev-Andrievskiy A, Popova A, Boyle R, Alberts J, Shenkman B et al (2014) Mice in Bion-M 1 space mission: training and selection. PLoS One 9(8):e104830. doi:10.1371/journal.pone. 0104830

Aracena P, Tang W, Hamilton SL, Hidalgo C (2005) Effects of S-glutathionylation and S-nitrosylation on calmodulin binding to triads and FKBP12 binding to type 1 calcium release channels. Antioxid Redox Signal 7:870–881

Arbeille P, Kerbeci P, Capri A, Dannaud C, Trappe SW, Trappe TA (2009) Quantification of muscle volume by echography: comparison with MRI data on subjects in long-term bed rest. Ultrasound Med Biol 35:1092–1097

Artioli GG, de Oliveira Silvestre JG, Guilherme JP, Baptista IL, Ramos GV, da Silva WJ, Miyabara EH, Moriscot AS (2014) Embryonic stem cells improve skeletal muscle recovery after extreme atrophy in mice. Muscle Nerve. doi:10.1002/mus.24320

Baldwin KM, Haddad F (2002) Skeletal muscle plasticity: cellular and molecular responses to altered physical activity paradigms. Am J Phys Med Rehabil 81:S40–S51

Ballard RW, Rossberg Walker K (1992) Flying U.S. science on the U.S.S.R. Cosmos biosatellites. ASGSB Bull 6:121–128

Balon TW, Nadler JL (1994) Nitric oxide release is present from incubated skeletal muscle preparations. J Appl Physiol 77:2519–2521

Belavý DL, Miokovic T, Armbrecht G, Rittweger J, Felsenberg D (2009) Resistive vibration exercise reduces lower limb muscle atrophy during 56-day bed-rest. J Musculoskelet Neuronal Interact 9:225–235

Belavý DL, Bock O, Börst H, Armbrecht G, Gast U, Degner C, Beller G, Soll H, Salanova M, Habazettl H, Heer M, de Haan A, Stegeman DF, Cerretelli P, Blottner D, Rittweger J, Gelfi C, Kornak U, Felsenberg D (2010) The 2nd Berlin BedRest study: protocol and implementation. J Musculoskelet Neuronal Interact 10:207–219

Belavý DL, Miokovic T, Rittweger J, Felsenberg D (2011) Estimation of changes in volume of individual lower-limb muscles using magnetic resonance imaging (during bed-rest). Physiol Meas 32(1):35–50

Belavý DL, Wilson SJ, Armbrecht G, Rittweger J, Felsenberg D, Richardson CA (2012) Resistive vibration exercise during bed-rest reduces motor control changes in the lumbo-pelvic musculature. J Electromyogr Kinesiol 22:21–30. doi:10.1016/j.jelekin.2011.09.009. Epub 2011 Oct 20

Bellinger AM, Reiken S, Dura M, Murphy PW, Deng SX, Landry DW, Nieman D, Lehnart SE, Samaru M, LaCampagne A, Marks AR (2008) Remodeling of ryanodine receptor complex causes "leaky" channels: a molecular mechanism for decreased exercise capacity. Proc Natl Acad Sci U S A 105(6):2198–2202

Bellinger AM, Reiken S, Carlson C, Mongillo M, Liu X, Rothman L, Matecki S, Lacampagne A, Marks AR (2009) Hypernitrosylated ryanodine receptor calcium release channels are leaky in dystrophic muscle. Nat Med 15(3):325–330. doi:10.1038/nm.1916. Epub 2009 Feb 8

Benavides Damm T, Franco-Obregón A, Egli M (2013) Gravitational force modulates G2/M phase exit in mechanically unloaded myoblasts. Cell Cycle 12:3001–3012

Benjamin M, Toumi H, Ralphs JR, Bydder G, Best TM, Milz S (2006) Where tendons and ligaments meet bone: attachment sites ('entheses') in relation to exercise and/or mechanical load. J Anat 208:471–490

Berg HE, Tedner B, Tesch PA (1993) Changes in lower limb muscle cross-sectional area and tissue fluid volume after transition from standing to supine. Acta Physiol Scand 148:379–385

Bergstrom J (1975) Percutaneous needle biopsy of skeletal muscle in physiological and clinical research. Scand J Clin Lab Invest 35:609–616

Blaauw B, Schiaffino S, Reggiani C (2013) Mechanisms modulating skeletal muscle phenotype. Compr Physiol 3(4):1645–1687. doi:10.1002/cphy.c130009

Blottner D (2013) Functional anatomy of skeletal muscle. In: Mueller-Wohlfarth HW et al (eds) Muscle injuries in sports, 2nd edn. Thieme Publishers, Stuttgart/New York

Blottner D, Lück G (1998) Nitric oxide synthase (NOS) in mouse skeletal muscle development and differentiated myoblasts. Cell Tissue Res 292:293–302

Blottner D, Lück G (2001) Just in time and place: NOS/NO system assembly in neuromuscular junction formation. Microsc Res Tech 55(3):171–180

Blottner D, Bosutti A, Degens H, Schiffl G, Gutsmann M, Buehlmeier J, Rittweger J, Ganse B, Heer M, Salanova M (in press) Whey protein plus bicarbonate has little effects on structural atrophy and proteolysis marker immunopatterns in skeletal muscle disuse during 21 days of bed rest. J Musculoskelet Neural Interact

Blottner D, Salanova M, Püttmann B, Schiffl G, Felsenberg D, Buehring B, Rittweger J (2006) Human skeletal muscle structure and function preserved by vibration muscle exercise following 55 days of bed rest. Eur J Appl Physiol 97:261–271

Blottner D, Serradj N, Salanova M, Touma C, Palme R, Silva M, Aerts JM, Berckmans D, Vico L, Liu Y, Giuliani A, Rustichelli F, Cancedda R, Jamon M (2009) Morphological, physiological and behavioural evaluation of a 'Mice in Space' housing system. J Comp Physiol B 179: 519–533

Bodine-Fowler SC, Pierotti DJ, Talmadge RJ (1995) Functional and cellular adaptation to weightlessness in primates. J Gravit Physiol 2:P43–P46

Bodine SC, Baar K (2012) Analysis of skeletal muscle hypertrophy in models of increased loading. Methods Mol Biol 798:213–229

Bogaerts A, Delecluse C, Claessens AL, Coudyzer W, Boonen S, Verschueren SM (2007) Impact of whole-body vibration training versus fitness training on muscle strength and muscle mass in older men: a 1-year randomized controlled trial. J Gerontol A Biol Sci Med Sci 62:630–635

Booth FW (1994) Terrestrial applications of bone and muscle research in microgravity. Adv Space Res 14:373–376

Bredt DS, Snyder SH (1990) Isolation of nitric oxide synthetase, a calmodulin-requiring enzyme. Proc Natl Acad Sci U S A 87:682–685

Brenman JE, Chao DS, Xia H, Aldape K, Bredt DS (1995) Nitric oxide synthase complexed with dystrophin and absent from skeletal muscle sarcolemma in Duchenne muscular dystrophy. Cell 82:743–752

Brenman JE, Xia H, Chao DS, Black SM, Bredt DS (1997) Regulation of neuronal nitric oxide synthase through alternative transcripts. Dev Neurosci 19:224–231

Brenner B, Maasen N (2013) Basic physiology and aspects of exercise. In: Mueller-Wohlfarth HW et al (eds) Muscle injuries in sports, 2nd edn. Thieme Publishers, Stuttgart

Buehring B, Belavy DL, Michaelis I, Gast U, Felsenberg D, Rittweger J (2011) Changes in lower extremity muscle function after 56 days of bed rest. J Appl Physiol 111(1):87–94

Camerino GM, Pierno S, Liantonio A, De Bellis M, Cannone M, Sblendorio V, Conte E, Mele A, Tricarico D, Tavella S, Ruggiu A, Cancedda R, Ohira Y, Danieli-Betto D, Ciciliot S, Germinario E, Sandonà D, Betto R, Camerino DC, Desaphy JF (2013) Effects of pleiotrophin overexpression on mouse skeletal muscles in normal loading and in actual and simulated microgravity. PLoS One 8:e72028

Cancedda R, Liu Y, Ruggiu A, Tavella S, Biticchi R, Santucci D, Schwartz S, Ciparelli P, Falcetti G, Tenconi C, Cotronei V, Pignataro S (2012) The Mice Drawer System (MDS) experiment and the space endurance record-breaking mice. PLoS One 7:e32243

Caron MA, Charette SJ, Maltais F, Debigaré R (2011) Variability of protein level and phosphorylation status caused by biopsy protocol design in human skeletal muscle analyses. BMC Res Notes 4:488. doi:10.1186/1756-0500-4-488

Casellato C, Tagliabue M, Pedrocchi A, Ferrigno G, Pozzo T (2007) How does microgravity affect the muscular and kinematic synergies in a complex movement? J Gravit Physiol 14(1): P93–P94

Chao DS, Gorospe JR, Brenman JE, Rafael JA, Peters MF, Froehner SC, Hoffman EP, Chamberlain JS, Bredt DS (1996) Selective loss of sarcolemmal nitric oxide synthase in Becker muscular dystrophy. J Exp Med 184:609–618

Chopard A, Hillock S, Jasmin BJ (2009) Molecular events and signalling pathways involved in skeletal muscle disuse-induced atrophy and the impact of countermeasures. J Cell Mol Med 13(9B):3032–3050. doi:10.1111/j.1582-4934.2009.00864.x. Epub 2009 Jul 28

Ciciliot S, Rossi AC, Dyar KA, Blaauw B, Schiaffino S (2013) Muscle type and fiber type specificity in muscle wasting. Int J Biochem Cell Biol 45(10):2191–2199. doi:10.1016/j.biocel.2013.05.016

Clark BC (2009) In vivo alterations in skeletal muscle form and function after disuse atrophy. Med Sci Sports Exerc 41:1869–1875

Cochrane DJ (2011) Vibration exercise: the potential benefits. Int J Sports Med 32:75–99

Convertino VA, Doerr DF, Stein SL (1989) Changes in size and compliance of the calf after 30 days of simulated microgravity. J Appl Physiol 66:1509–1512

Crosbie RH, Barresi R, Campbell KP (2002) Loss of sarcolemma nNOS in sarcoglycan-deficient muscle. FASEB J 16:1786–1791

Davidson JM, Aquino AM, Woodward SC, Wilfinger WW (1999) Sustained microgravity reduces intrinsic wound healing and growth factor responses in the rat. FASEB J 13(2):325–329

Davis SA, Davis BL (2012) Exercise equipment used in microgravity: challenges and opportunities. Curr Sports Med Rep 11:142–147

Denise P, Besnard S, Vignaux G, Sabatier JP, Edy E, Hitier M, Levasseur R (2009) Sympathetic B antagonist prevents bone mineral density decrease induced by labyrinthectomy. Aviakosm Ekolog Med 43:36–38

Desplanches D (1997) Structural and functional adaptations of skeletal muscle to weightlessness. Int J Sports Med 18(Suppl 4):S259–S264

Dowling P, Holland A, Ohlendieck K (2014) Mass spectrometry-based identification of muscle-associated and muscle-derived proteomic biomarkers of dystrophinopathies. J Neuromusc Dis. doi:10.3233/JND-140011

Duke PJ, Montufar-Solis D (1999) Exposure to altered gravity affects all stages of endochondral cartilage differentiation. Adv Space Res 24:821–827

Du J, Wang X, Miereles C, Bailey JL, Debigare R, Zheng B, Price SR, Mitch WE (2004) Activation of caspase-3 is an initial step triggering accelerated muscle proteolysis in catabolic conditions. J Clin Invest 113:115–123

Eckberg DL, Neurolab Autonomic Nervous System Team (2003) Bursting into space: alterations of sympathetic control by space travel. Acta Physiol Scand 177:299–311

Edgerton VR, McCall GE, Hodgson JA, Gotto J, Goulet C, Fleischmann K, Roy RR (2001) Sensorimotor adaptations to microgravity in humans. J Exp Biol 204:3216–3223

English C, Fisher L, Thoirs K (2012) Reliability of real-time ultrasound for measuring skeletal muscle size in human limbs in vivo: a systematic review. Clin Rehabil 26:934–944

Ertl AC, Diedrich A, Biaggioni I, Levine BD, Robertson RM, Cox JF, Zuckerman JH, Pawelczyk JA, Ray CA, Buckey JC Jr, Lane LD, Shiavi R, Gaffney FA, Costa F, Holt C, Blomqvist CG, Eckberg DL, Baisch FJ, Robertson D (2002) Human muscle sympathetic nerve activity and plasma noradrenaline kinetics in space. J Physiol 538(Pt 1):321–329

Esformes JI, Narici MV, Maganaris CN (2002) Measurement of human muscle volume using ultrasonography. Eur J Appl Physiol 87:90–92

Eu JP, Hare JM, Hess DT, Skaf M, Sun J, Cardenas-Navina I, Sun QA, Dewhirst M, Meissner G, Stamler JS (2003) Concerted regulation of skeletal muscle contractility by oxygen tension and endogenous nitric oxide. Proc Natl Acad Sci USA 100(25):15229–15234. Epub 2003 Nov 26

Fanin M, Tasca E, Nascimbeni AC, Angelini C (2009) Sarcolemmal neuronal nitric oxide synthase defect in limb-girdle muscular dystrophy: an adverse modulating factor in the disease course? J Neuropathol Exp Neurol 68:383–390

Faulkner JA (2003) Terminology for contractions of muscles during shortening, while isometric, and during lengthening. J Appl Physiol 95:455–459

Ferreira JA, Crissey JM, Brown M (2011) An alternant method to the traditional NASA hindlimb unloading model in mice. J Vis Exp 10(49). pii: 2467.

Fitts RH, Riley DR, Widrick JJ (2001) Functional and structural adaptations of skeletal muscle to microgravity. J Exp Biol 204:3201–3208

Fitts RH, Trappe SW, Costill DL, Gallagher PM, Creer AC, Colloton PA, Peters JR, Romatowski JG, Bain JL, Riley DA (2010) Prolonged space flight-induced alterations in the structure and function of human skeletal muscle fibres. J Physiol 588:3567–3592

Flück M (2006) Functional, structural and molecular plasticity of mammalian skeletal muscle in response to exercise stimuli. J Exp Biol 209:2239–2248

Foster MW, Liu L, Zeng M, Hess DT, Stamler JS (2009) A genetic analysis of nitrosative stress. Biochemistry 48:792–799

Frandsen U, Höffner L, Betak A, Saltin B, Bangsbo J, Hellsten Y (2000) Endurance training does not alter the level of neuronal nitric oxide synthase in human skeletal muscle. J Appl Physiol 89:1033–1038

Franzoso S, Sandona D, Picard A, Furlan S, Gobbo V, Salvatori S, Elvassore N, Cimetta E, Betto R, Schiaffino S (2009) Cultured adult muscle fibers in the microgravity environment. The MYO experiment in the Foton-M3 space flight mission. Basis Appl Myol 19:65–76

Fujii Y, Guo Y, Hussain SN (1998) Regulation of nitric oxide production in response to skeletal muscle activation. J Appl Physiol 85:2330–2336

Gath I, Ebert J, Gödtel-Armbrust U, Ross R, Reske-Kunz AB, Förstermann U (1999) NO synthase II in mouse skeletal muscle is associated with caveolin 3. Biochem J 340:723–728

Gelfi C, Vasso M, Cerretelli P (2011) Diversity of human skeletal muscle in health and disease: contribution of proteomics. J Proteomics 74:774–795

Genc KO, Gopalakrishnan R, Kuklis MM, Maender CC, Rice AJ, Bowersox KD, Cavanagh PR (2010) Foot forces during exercise on the International Space Station. J Biomech 43: 3020–3027

Gopalakrishnan R, Genc KO, Rice AJ, Lee SM, Evans HJ, Maender CC, Ilaslan H, Cavanagh PR (2009) Muscle volume, strength, endurance, and exercise loads during 6-month missions in space. Aviat Space Environ Med 81:91–102

Grozdanovic Z, Baumgarten HG (1999) Nitric oxide synthase in skeletal muscle fibers: a signaling component of the dystrophin-glycoprotein complex. Histol Histopathol 14:243–256, Review

Handschin C (2010) Regulation of skeletal muscle cell plasticity by the peroxisome proliferator-activated receptor γ coactivator 1α. J Recept Signal Transduct Res 30:376–384. doi:10.3109/10799891003641074. Epub 2010 Feb 24

Hargens AR, Bhattacharya R, Schneider SM (2013) Space physiology VI: exercise, artificial gravity, and countermeasure development for prolonged space flight. Eur J Appl Physiol 113(9):2183–2192. doi:10.1007/s00421-012-2523-5

Hellweg CE, Baumstark-Khan C (2007) Getting ready for the manned mission to Mars: the astronauts' risk from space radiation. Naturwissenschaften 94:517–526

Hess DT, Stamler JS (2011) Regulation by S-nitrosylation of protein post-translational modification. J Biol Chem 287(7):4411–4418

Hides JA, Belavý DL, Stanton W, Wilson SJ, Rittweger J, Felsenberg D, Richardson CA (2007) Magnetic resonance imaging assessment of trunk muscles during prolonged bed rest. Spine 32:1687–1692

Hoh JF (2005) Laryngeal muscle fibre types. Acta Physiol Scand 183:133–149

Honaker JA1, Samy RN (2007) Vestibular-evoked myogenic potentials. Curr Opin Otolaryngol Head Neck Surg 15:330–334

Hughes-Fulford M (2003) Function of the cytoskeleton in gravisensing during spaceflight. Adv Space Res 32:1585–1593

Hughes-Fulford M (2004) Signal transduction and mechanical stress. Sci STKE 2004(249):RE12

Huijing PA, Baan GC (2003) Myofascial force transmission: muscle relative position and length determine agonist and synergist muscle force. J Appl Physiol 94:1092–1107

Hussain SN, El-Dwairi Q, Abdul-Hussain MN, Sakkal D (1997) Expression of nitric oxide synthase isoforms in normal ventilatory and limb muscles. J Appl Physiol 83:348–353

Ito N, Ruegg UT, Kudo A, Miyagoe-Suzuki Y, Takeda S (2013) Activation of calcium signaling through Trpv1 by nNOS and peroxynitrite as a key trigger of skeletal muscle hypertrophy. Nat Med 19:101–106

Ilyin EA (2000) Historical overview of the Bion project. J Gravit Physiol 7:S1–S8

Ilyin EA (2007) From the first dog to the last monkey in space. J Gravit Physiol 14:P143–P146

Jackman RW, Kandarian SC (2004) The molecular basis of skeletal muscle atrophy. Am J Physiol Cell Physiol 287:C834–C843

Jaffrey SR, Snyder SH (2001) The biotin switch method for the detection of S-nitrosylated proteins. Sci STKE 2001(86):pl1

Jaffrey SR, Erdjument-Bromage H, Ferris CD, Tempst P, Snyder SH (2001) Protein S-nitrosylation: a physiological signal for neuronal nitric oxide. Nat Cell Biol 3:193–197

Jamon M (2014) The development of vestibular system and related functions in mammals: impact of gravity. Front Integr Neurosci 8:11. doi:10.3389/fnint.2014.00011

Ji LL (2008) Modulation of skeletal muscle antioxidant defense by exercise: role of redox signaling. Free Radic Biol Med 44:142–152

Kasri M, Picquet F, Falempin M (2004) Effects of unilateral and bilateral labyrinthectomy on rat postural muscle properties: the soleus. Exp Neurol 185:143–153

Katahira K (2001) Global harmonization in the care and use of laboratory animals for the space life science research. Biol Sci Space 15(1):30–34

Kingwell BA (2000) Nitric oxide as a metabolic regulator during exercise: effects of training in health and disease. Clin Exp Pharmacol Physiol 27:239–250

Kobayashi YM, Rader EP, Crawford RW, Iyengar NK, Thedens DR, Faulkner JA, Parikh SV, Weiss RM, Chamberlain JS, Moore SA, Campbell KP (2008) Sarcolemma-localized nNOS is required to maintain activity after mild exercise. Nature 456:511–515

Koryak Y (2002) DRY immersion induces neural and contractile adaptations in the human triceps surae muscle. Environ Med 46(1–2):17–27

Kramer A, Gollhofer A, Ritzmann R (2013) Acute exposure to microgravity does not influence the H-reflex with or without whole body vibration and does not cause vibration-specific changes in muscular activity. J Electromyogr Kinesiol 23:872–878

Kusner LL, Kaminski HJ (1996) Nitric oxide synthase is concentrated at the skeletal muscle endplate. Brain Res 730(1–2):238–242

Lackner JR, Dizio P (2006) Space motion sickness. Exp Brain Res 17:377–399

Larina IM, Tcheglova IA, Shenkman BS, Nemirovskaya TL (1997) Muscle atrophy and hormonal regulation in women in 120 day bed rest. J Gravit Physiol 4:P121–P122

Lawler JM, Song W, Demaree SR (2003) Hindlimb unloading increases oxidative stress and disrupts antioxidant capacity in skeletal muscle. Free Radic Biol Med 35:9–16

Lawler JM, Kunst M, Jeff M, Hord JM, Lee Y, Joshi K, Botchlett RE, Ramirez A, Martinez DA (2014) EUK-134 ameliorates nNOS translocation and skeletal muscle fiber atrophy during short-term mechanical unloading. Am J Physiol Regul Integr Comp Physiol 306:R470–R482

LeBlanc AD, Schneider VS, Evans HJ, Pientok C, Rowe R, Spector E (1992) Regional changes in muscle mass following 17 weeks of bed rest. J Appl Physiol 73:2172–2178

LeBlanc A, Rowe R, Evans H, West S, Shackelford L, Schneider V (1997) Muscle atrophy during long duration bed rest. Int J Sports Med 18(Suppl 4):S283–S285

LeBlanc AD, Spector ER, Evans HJ, Sibonga JD (2007) Skeletal responses to space flight and the bed rest analog: a review. J Musculoskelet Neuronal Interact 7(1):33–47

Lehnart SE, Wehrens XH, Reiken S, Warrier S, Belevych AE, Harvey RD, Richter W, Jin SL, Conti M, Marks AR (2005) Phosphodiesterase 4D deficiency in the ryanodine-receptor complex promotes heart failure and arrhythmias. Cell 123(1):25–35

Li D, Shin JH, Duan D (2011) iNOS ablation does not improve specific force of the extensor digitorum longus muscle in dystrophin-deficient mdx4cv mice. PLoS One 6:e21618

Li R, Narici MV, Erskine RM, Seynnes OR, Rittweger J, Pišot R, Šimunič B, Flück M (2013) Costamere remodeling with muscle loading and unloading in healthy young men. J Anat 223(5):525–536. doi:10.1111/joa.12101

Lima B, Forrester MT, Hess DT, Stamler JS (2010) S-nitrosylation in cardiovascular signaling. Circ Res 106:633–646

Lück G, Oberbäumer I, Blottner D (1998) In situ identification of neuronal nitric oxide synthase (NOS-I) mRNA in mouse and rat skeletal muscle. Neurosci Lett 246:77–80

Lück G, Hoch W, Hopf C, Blottner D (2000) Nitric oxide synthase (NOS-1) coclustered with agrin-induced AChR-specializations on cultured skeletal myotubes. Mol Cell Neurosci 16 (3):269–281

Luxa N, Salanova M, Schiffl G, Gutsmann M, Besnard S, Denise P, Clarke A, Blottner D (2013) Increased myofiber remodelling and NFATc1-myonuclear translocation in rat postural skeletal muscle after experimental vestibular deafferentation. J Vestib Res 23:187–193

Macias BR, Groppo ER, Eastlack RK, Watenpaugh DE, Lee SM, Schneider SM, Boda WL, Smith SM, Cutuk A, Pedowitz RA, Meyer RS, Hargens AR (2005) Space exercise and earth benefits. Curr Pharm Biotechnol 6:305–317. LBMP as countermeasure

Mano T (2005) Autonomic neural functions in space. Curr Pharm Biotechnol 6:319–324

Mano T, Iwase S (2003) Sympathetic nerve activity in hypotension and orthostatic intolerance. Acta Physiol Scand 177:359–365

Martins KJ, St-Louis M, Murdoch GK, MacLean IM, McDonald P, Dixon WT, Putman CT, Michel RN (2012) Nitric oxide synthase inhibition prevents activity-induced calcineurin-NFATc1 signalling and fast-to-slow skeletal muscle fibre type conversions. J Physiol 590:1427–1442

Masi AT, Nair K, Evans T, Ghandour Y (2010) Clinical, biomechanical, and physiological translational interpretations of human resting myofascial tone or tension. Int J Ther Massage Bodywork 16:16–28

Mayet-Sornay MH, Hoppeler H, Shenkman BS, Desplanches D (2000) Structural changes in arm muscles after microgravity. J Gravit Physiol 7:S43–S44

McKeehen JN, Novotny SA, Baltgalvis KA, Call JA, Nuckley DJ, Lowe DA (2013) Adaptations of mouse skeletal muscle to low-intensity vibration training. Med Sci Sports Exerc 45(6): 1051–1059. doi:10.1249/MSS.0b013e3182811947

Meinen S, Lin S, Rüegg MA, Punga AR (2012) Fatigue and muscle atrophy in a mouse model of myasthenia gravis is paralleled by loss of sarcolemmal nNOS. PLoS One 7:e44148

Mester J, Kleinöder H, Yue Z (2006) Vibration training: benefits and risks. J Biomech 39: 1056–1065

Miokovic T, Armbrecht G, Felsenberg D, Belavy DL (2011) Differential atrophy of the postero-lateral hip musculature during prolonged bed rest and the influence of exercise countermeasures. J Appl Physiol 110:926–934. doi:10.1152/japplphysiol011052010. Epub 2011 Jan 13

Miokovic T, Armbrecht G, Felsenberg D, Belavý DL (2012) Heterogeneous atrophy occurs within individual lower limb muscles during 60 days of bed rest. J Appl Physiol 113:1545–1559. doi:10.1152/japplphysiol006112012. Epub 2012 Sep 13

Miyagoe-Suzuki Y, Takeda SI (2001) Association of neuronal nitric oxide synthase (nNOS) with alpha1-syntrophin at the sarcolemma. Microsc Res Tech 55:164–170

Momken I, Stevens L, Bergouignan A, Desplanches D, Rudwill F, Chery I, Zahariev A, Zahn S, Stein TP, Sebedio JL, Pujos-Guillot E, Falempin M, Simon C, Coxam V, Andrianjafiniony T, Gauquelin-Koch G, Picquet F, Blanc S (2011) Resveratrol prevents the wasting disorders of mechanical unloading by acting as a physical exercise mimetic in the rat. FASEB J 25: 3646–3660

Morey-Holton ER, Globus RK (2002) Hindlimb unloading rodent model: technical aspects. J Appl Physiol 92:1367–1377

Morey-Holton ER, Hill EL, Souza KA (2007) Animals and spaceflight: from survival to under-standing. J Musculoskelet Neuronal Interact 7:17–25

Moriggi M, Vasso M, Fania C, Capitanio D, Bonifacio G, Salanova M, Blottner D, Rittweger J, Felsenberg D, Cerretelli P, Gelfi C (2010) Long term bed rest with and without vibration exercise countermeasures: effects on human muscle protein dysregulation. Proteomics 10:3756–3774

Moukhina A, Shenkman B, Blottner D, Nemirovskaya T, Lemesheva Y, Püttmann B, Kozlovskaya I (2004) Effects of support stimulation on human soleus fiber characteristics during exposure to "dry" immersion. J Gravit Physiol 11:P137–P138

Mulder ER, Stegeman DF, Gerrits KH, Paalman MI, Rittweger J, Felsenberg D, de Haan A (2006) Strength, size and activation of knee extensors followed during 8 weeks of horizontal bed rest and the influence of a countermeasure. Eur J Appl Physiol 97:706–715

Murton AJ, Constantin D, Greenhaff PL (2008) The involvement of the ubiquitin proteasome system in human skeletal muscle remodelling and atrophy. Biochim Biophys Acta 1782: 730–743

Nakane M, Schmidt HH, Pollock JS, Förstermann U, Murad F (1993) Cloned human brain nitric oxide synthase is highly expressed in skeletal muscle. FEBS Lett 316:175–180

Narici MV, de Boer MD (2011) Disuse of the musculo-skeletal system in space and on earth. Eur J Appl Physiol 111:403–420

National Research Council NRC (1996) Guide for the care and use of laboratory animals. National Academy Press, Washington, DC, p 124

Nedergaard A, Sun S, Karsdal MA, Henriksen K, Kjær M, Lou Y, He Y, Zheng Q, Suetta C (2013) Type VI collagen turnover-related peptides-novel serological biomarkers of muscle mass and anabolic response to loading in young men. J Cachexia Sarcopenia Muscle 4:267–275

Nikolaidis MG, Kyparos A, Spanou C, Paschalis V, Theodorou AA, Vrabas IS (2012) Redox biology of exercise: an integrative and comparative consideration of some overlooked issues. J Exp Biol 215(Pt 10):1615–1625. doi:10.1242/jeb.067470, PMID 22539728

Nogueira L, Figueiredo-Freitas C, Casimiro-Lopes G, Magdesian MH, Assreuy J, Sorenson MM (2009) Myosin is reversibly inhibited by S-nitrosylation. Biochem J 424:221–231

Oganov VS, Potapov AN (2006) Functional plasticity of mammalian skeletal muscles under microgravity. Aviakosm Ekolog Med 40(1):27–35

Ohira Y, Jiang B, Roy RR, Oganov V, Ilyina-Kakueva E, Marini JF, Edgerton VR (1992) Rat soleus muscle fiber responses to 14 days of spaceflight and hindlimb suspension. J Appl Physiol 73:51S–57S

Ohira Y, Yoshinaga T, Nomura T, Kawano F, Ishihara A, Nonaka I, Roy RR, Edgerton VR (2002) Gravitational unloading effects on muscle fiber size, phenotype and myonuclear number. Adv Space Res 30:777–781

Ohira T, Ohira T, Kawano F, Shibaguchi T, Okabe H, Goto K, Ogita F, Sudoh M, Roy RR, Edgerton VR, Cancedda R, Ohira Y (2014) Effects of gravitational loading levels on protein expression related to metabolic and/or morphologic properties of mouse neck muscles. Physiol Rep 2:e00183

Olesen J, Kiilerich K, Pilegaard H (2010) PGC-1alpha-mediated adaptations in skeletal muscle. Pflugers Arch 460(1):153–162. doi:10.1007/s00424-010-0834-0. Epub 2010 Apr 19

Oliver L, Goureau O, Courtois Y, Vigny M (1996) Accumulation of NO synthase (type-I) at the neuromuscular junctions in adult mice. Neuroreport 7:924–926

Paoli A, Pacelli QF, Toniolo L, Miotti D, Reggiani C (2010) Latissimus dorsi fine needle muscle biopsy: a novel and efficient approach to study proximal muscles of upper limbs. J Surg Res 164:e257–e263. doi:10.1016/j.jss.2010.05.043. Epub 2010 Jun 12

Pavy-Le Traon A, Heer M, Narici MV, Rittweger J, Vernikos J (2006) From space to Earth: advances in human physiology from 20 years of bed rest studies (1986–2006). Eur J Appl Physiol 101:143–194

Percival JM, Anderson KN, Gregorevic P, Chamberlain JS, Froehner SC (2008) Functional deficits in nNOSmu-deficient skeletal muscle: myopathy in nNOS knockout mice. PLoS One 10:e3387

Percival JM, Anderson KN, Huang P, Adams ME, Froehner SC (2010) Golgi and sarcolemmal neuronal NOS differentially regulate contraction-induced fatigue and vasoconstriction in exercising mouse skeletal muscle. J Clin Invest 120:816–826

Peterson JM, Bakkar N, Guttridge DC (2011) NF-κB signaling in skeletal muscle health and disease. Curr Top Dev Biol 96:85–119. doi:10.1016/B978-0-12-385940-2.00004-8

Planitzer G, Baum O, Gossrau R (2000) Skeletal muscle fibres show NADPH diaphorase activity associated with mitochondria, the sarcoplasmic reticulum and the NOS-1-containing sarcolemma. Histochem J 32:303–312

Ploutz-Snyder LL, Nyren S, Cooper TG, Potchen EJ, Meyer RA (1997) Different effects of exercise and edema on T2 relaxation in skeletal muscle. Magn Reson Med 37:676–682

Powers SK, Lennon SL (1999) Analysis of cellular responses to free radicals: focus on exercise and skeletal muscle. Proc Nutr Soc 58:1025–1033, Review

Powers SK, Kavazis AN, McClung JM (2007) Oxidative stress and disuse muscle atrophy. J Appl Physiol 102:2389–2397

Punkt K, Schering S, Fritzsche M, Asmussen G, Minin EA, Samoilova VE, Müller FU, Schmitz W, Hasselblatt M, Paulus W, Müller-Werdan U, Slezak J, Koehler G, Boecker W, Buchwalow IB (2007) Fibre-related nitric oxide synthase (NOS) in Duchenne muscular dystrophy. Acta Histochem 109:228–236

Püttmann B, Gerlach EM, Krüger M, Blottner D (2005) Neuromuscular contacts induce nitric oxide signals in skeletal myotubes in vitro. Neurosignals 14:85–95

Rabin R, Gordon SL, Lymn RW, Todd PW, Frey MA, Sulzman FM (1993) Effects of spaceflight on the musculoskeletal system: NIH and NASA future directions. FASEB J 7:396–398

Rauch F, Sievanen H, Boonen S, Cardinale M, Degens H, Felsenberg D, Roth J, Schoenau E, Verschueren S, Rittweger J; International Society of Musculoskeletal and Neuronal Interactions (2010) Reporting whole-body vibration intervention studies: recommendations of the International Society of Musculoskeletal and Neuronal Interactions. J Musculoskelet Neuronal Interact 10:193–198

Reiser PJ, Kline WO, Vaghy PL (1997) Induction of neuronal type nitric oxide synthase in skeletal muscle by chronic electrical stimulation in vivo. J Appl Physiol 82:1250–1255

Reschke MF, Bloomberg JJ, Paloski WH, Mulavara AP, Feiveson AH, Harm DL (2009) Postural reflexes, balance control, and functional mobility with long-duration head-down bed rest. Aviat Space Environ Med 80(5 Suppl):A45–A54

Riley D (1999) Is skeletal muscle ready for long-term spaceflight and return to gravity? Adv Space Biol Med 7:31–48

Riley DA, Bain JL, Thompson JL, Fitts RH, Widrick JJ, Trappe SW, Trappe TA, Costill DL (2002) Thin filament diversity and physiological properties of fast and slow fiber types in astronaut leg muscles. J Appl Physiol 92:817–225

Rittweger J, Belavy D, Hunek P, Gast U, Boerst H, Feilcke B, Armbrecht G, Mulder E, Schubert H, Richardson C, de Haan A, Stegeman DF, Schiessl H, Felsenberg D (2006) Highly demanding resistive vibration exercise program is tolerated during 56 days of strict bed-rest. Int J Sports Med 27:553–559

Rittweger J (2010) Vibration as an exercise modality: how it may work, and what its potential might be. Eur J Appl Physiol 108:877–904

Rittweger J, Möller K, Bareille MP, Felsenberg D, Zange J (2013) Muscle X-ray attenuation is not decreased during experimental bed rest. Muscle Nerve 47:722–730

Ritzmann R, Kramer A, Bernhardt S, Gollhofer A (2014) Whole body vibration training–improving balance control and muscle endurance. PLoS One 26:e89905

Rizzo AM, Corsetto PA, Montorfano G, Milani S, Zava S, Tavella S, Cancedda R, Berra B (2012) Effects of long-term space flight on erythrocytes and oxidative stress of rodents. PLoS One 7: e32361

Roy RR, Zhong H, Talmadge RJ, Bodine SC, Fanton JW, Koslovskaya I, Edgerton VR (2000) Size and myonuclear domains in Rhesus soleus muscle fibers: short-term spaceflight. J Gravit Physiol 7:S45

Rubinstein I, Abassi Z, Coleman R, Milman F, Winaver J, Better OS (1998) Involvement of nitric oxide system in experimental muscle crush injury. J Clin Invest 101:1325–1333

Rudnick J, Püttmann B, Tesch PA, Alkner B, Schoser BG, Salanova M, Kirsch K, Gunga HC, Schiffl G, Lück G, Blottner D (2004) Differential expression of nitric oxide synthases (NOS 1–3) in human skeletal muscle following exercise countermeasure during 12 weeks of bed rest. FASEB J 18:1228–1230

Salanova M, Schiffl G, Püttmann B, Schoser BG, Blottner D (2008) Molecular biomarkers monitoring human skeletal muscle fibres and microvasculature following long-term bed rest with and without countermeasures. J Anat 212:306–318

Salanova M, Schiffl G, Rittweger J, Felsenberg D, Blottner D (2009a) Ryanodine receptor type-1 (RyR1) expression and protein S-nitrosylation pattern in human soleus myofibres following bed rest and exercise countermeasure. Histochem Cell Biol 130:105–118

Salanova M, Schiffl G, Blottner D (2009b) Atypical fast SERCA1a protein expression in slow myofibers and differential S-nitrosylation prevented by exercise during long term bed rest. Histochem Cell Biol 132:383–394. doi:10.1007/s00418-009-0624-y

Salanova M, Bortoloso E, Schiffl G, Gutsmann M, Belavy DL, Felsenberg D, Furlan S, Volpe P, Blottner D (2011) Expression and regulation of Homer in human skeletal muscle during neuromuscular junction adaptation to disuse and exercise. FASEB J 25:4312–4325. doi:10.1096/fj.11-186049

Salanova M, Schiffl G, Gutsmann M, Felsenberg D, Furlan S, Volpe P, Clarke A, Blottner D (2013) Nitrosative stress in human skeletal muscle attenuated by exercise countermeasure after chronic disuse. Redox Biol 1:514–526. doi:10.1016/j.redox.2013.10.006

Salanova M, Gelfi C, Moriggi M, Vasso M, Viganò A, Minafra L, Bonifacio G, Schiffl G, Gutsmann M, Felsenberg D, Cerretelli P, Blottner D (2014) Disuse deterioration of human skeletal muscle challenged by resistive exercise superimposed with vibration: evidence from structural and proteomic analysis. FASEB J 28:4748–4763

Sandonà D, Desaphy JF, Camerino GM, Bianchini E, Ciciliot S, Danieli-Betto D, Dobrowolny G, Furlan S, Germinario E, Goto K, Gutsmann M, Kawano F, Nakai N, Ohira T, Ohno Y, Picard A, Salanova M, Schiffl G, Blottner D, Musarò A, Ohira Y, Betto R, Conte D, Schiaffino S (2012) Adaptation of mouse skeletal muscle to long-term microgravity in the MDS mission. PLoS One 7(3):e33232

Santucci D, Kawano F, Ohira T, Terada M, Nakai N, Francia N, Alleva E, Aloe L, Ochiai T, Cancedda R, Goto K, Ohira Y (2012) Evaluation of gene, protein and neurotrophin expression in the brain of mice exposed to space environment for 91 days. PLoS One 7(7):e40112

Sarkis J, Vié V, Winder SJ, Renault A, Le Rumeur E, Hubert JF (2013) Resisting sarcolemmal rupture: dystrophin repeats increase membrane-actin stiffness. FASEB J 27:359–367

Schiaffino S, Reggiani C (2011) Fiber types in mammalian skeletal muscles. Physiol Rev 91(4): 1447–1531. doi:10.1152/physrev.00031.2010

Schleip R, Jäger H, Klingler W (2012) What is 'fascia'? A review of different nomenclatures. J Bodyw Mov Ther 16:496–502

Schmidt MA, Goodwin TJ (2013) Personalized medicine in human space flight: using Omics based analyses to develop individualized countermeasures that enhance astronaut safety and performance. Metabolomics 9:1134–1156

Schoenfeld BJ (2012) Does exercise-induced muscle damage play a role in skeletal muscle hypertrophy? J Strength Cond Res 26:1441–1453. doi:10.1519/JSC.0b013e31824f207e

Schoenfeld MP, Ansari RR, Zakrajsek JF, Billiar TR, Toyoda Y, Wink DA, Nakao A (2011) Hydrogen therapy may reduce the risks related to radiation-induced oxidative stress in space flight. Med Hypotheses 76:117–118. (Radiation induced oxidative damage reduced by inhalation of hydrogen)

Silvagno F, Xia H, Bredt DS (1996) Neuronal nitric-oxide synthase-mu, an alternatively spliced isoform expressed in differentiated skeletal muscle. J Biol Chem 271(19):11204–11208

Sitjà Rabert M, Rigau Comas D, Fort Vanmeerhaeghe A, Santoyo Medina C, Roqué i Figuls M, Romero-Rodríguez D, Bonfill Cosp X (2012) Whole-body vibration training for patients with neurodegenerative disease. Cochrane Database Syst Rev 2:CD009097

Seth D, Stamler JS (2011) The SNO-proteome: causation and classifications. Curr Opin Chem Biol 15:129–136

Smith LW, Smith JD, Criswell DS (2002) Involvement of nitric oxide synthase in skeletal muscle adaptation to chronic overload. J Appl Physiol 92:2005–2011

Stamler JS, Meissner G (2001) Physiology of nitric oxide in skeletal muscle. Physiol Rev 81: 209–237

Stein TP (2002) Space flight and oxidative stress. Nutrition 18:867–871

Stein TP, Leskiw MJ (2000) Oxidant damage during and after spaceflight. Am J Physiol Endocrinol Metab 278:E375–E382

Stein TP, Blanc S (2011) Does protein supplementation prevent muscle disuse atrophy and loss of strength? Crit Rev Food Sci Nutr 51:828–834

Suhr F, Gehlert S, Grau M, Bloch W (2013) Skeletal muscle function during exercise-fine-tuning of diverse subsystems by nitric oxide. Int J Mol Sci 14:7109–7139

Sun LW, Blottner D, Luan HQ, Salanova M, Wang C, Niu HJ, Felsenberg D, Fan YB (2013) Bone and muscle structure and quality preserved by active versus passive muscle exercise on a new stepper device in 21 days tail-suspended rats. J Musculoskelet Neuronal Interact 13:166–177

Suzuki N, Motohashi N, Uezumi A, Fukada S, Yoshimura T, Itoyama Y, Aoki M, Miyagoe-Suzuki Y, Takeda S (2007) NO production results in suspension-induced muscle atrophy through dislocation of neuronal NOS. J Clin Invest 117:2468–2476

Suzuki N, Mizuno H, Warita H, Takeda S, Itoyama Y, Aoki M (2010) Neuronal NOS is dislocated during muscle atrophy in amyotrophic lateral sclerosis. J Neurol Sci 294:95–101

Tarnopolsky MA, Pearce E, Smith K, Lach B (2011) Suction-modified Bergström muscle biopsy technique: experience with 13,500 procedures. Muscle Nerve 43:717–725. doi:10.1002/mus.21945

Tavella S, Ruggiu A, Giuliani A, Brun F, Canciani B, Manescu A, Marozzi K, Cilli M, Costa D, Liu Y, Piccardi F, Tasso R, Tromba G, Rustichelli F, Cancedda R (2012) Bone turnover in wild type and pleiotrophin-transgenic mice housed for three months in the International Space Station (ISS). PLoS One 7(3):e33179

Tengan CH, Rodrigues GS, Godinho RO (2012) Nitric oxide in skeletal muscle: role on mitochondrial biogenesis and function. Int J Mol Sci 13:17160–17184

Tews DS, Goebel HH, Schneider I, Gunkel A, Stennert E, Neiss WF (1997) Expression of different isoforms of nitric oxide synthase in experimentally denervated and reinnervated skeletal muscle. J Neuropathol Exp Neurol 56:1283–1289

Thornton WE, Hedge V, Coleman E, Uri JJ, Moore TP (1992) Changes in leg volume during microgravity simulation. Aviat Space Environ Med 63:789–794

Thyfault JP, Booth FW (2011) Lack of regular physical exercise or too much inactivity. Curr Opin Clin Nutr Metab Care 14:374–378

Tidball JG, Lavergne E, Lau KS, Spencer MJ, Stull JT, Wehling M (1998) Mechanical loading regulates NOS expression and activity in developing and adult skeletal muscle. Am J Physiol 275:C260–C266

Tidball JG, Wehling-Henricks M (2004) Expression of a NOS transgene in dystrophin-deficient muscle reduces muscle membrane damage without increasing the expression of membrane-associated cytoskeletal proteins. Mol Genet Metab 82:312–320

Tonon C, Gramegna LL, Lodi R (2012) Magnetic resonance imaging and spectroscopy in the evaluation of neuromuscular disorders and fatigue. Neuromuscul Disord 22(Suppl 3):S187–S191. doi:10.1016/j.nmd.2012.10.008

Tothova J, Blaauw B, Pallafacchina G, Rudolf R, Argentini C, Reggiani C, Schiaffino S (2006) NFATc1 nucleocytoplasmic shuttling is controlled by nerve activity in skeletal muscle. J Cell Sci 119:1604–1611

Trappe S, Costill D, Gallagher P, Creer A, Peters JR, Evans H, Riley DA, Fitts RH (2009) Exercise in space: human skeletal muscle after 6 months aboard the International Space Station. J Appl Physiol 106:1159–1168

Trappe TA, Standley RA, Liu SZ, Jemiolo B, Trappe SW, Harber MP (2013) Local anesthetic effects on gene transcription in human skeletal muscle biopsies. Muscle Nerve 48:591–593. doi:10.1002/mus.23860. Epub 2013 Aug 30

Tudor-Locke C, Craig CL, Brown WJ, Clemes SA, De Cocker K, Giles-Corti B, Hatano Y, Inoue S, Matsudo SM, Mutrie N, Oppert JM, Rowe DA, Schmidt MD, Schofield GM, Spence JC, Teixeira PJ, Tully MA, Blair SN (2011) How many steps/day are enough? For adults. Int J Behav Nutr Phys Act 8:79. doi:10.1186/1479-5868-8-79

Unsworth BR, Lelkes PI (1998) Growing tissues in microgravity. Nat Med 4:901–907

Urso ML (2009) Regulation of muscle atrophy: wasting away from the outside in: an introduction. Med Sci Sports Exerc 41:1856–1859

Vignaux G, Besnard S, Ndong J, Philoxène B, Denise P, Elefteriou F (2013) Bone remodeling is regulated by inner ear vestibular signals. J Bone Miner Res 28:2136–2144

Vitadello M, Gherardini J, Gorza L (2014) The stress protein/chaperone Grp94 counteracts muscle disuse atrophy by stabilizing subsarcolemmal neuronal nitric oxide synthase. Antioxid Redox Signal 20:2479–2496

Wang MX, Murrell DF, Szabo C, Warren RF, Sarris M, Murrell GA (2001) Nitric oxide in skeletal muscle: inhibition of nitric oxide synthase inhibits walking speed in rats. Nitric Oxide 5:219–232

Wang YY, Lin SY, Chuang YH, Mao CH, Tung KC, Sheu WH (2010) Protein nitration is associated with increased proteolysis in skeletal muscle of bile duct ligation-induced cirrhotic rats. Metabolism 59(4):468–472. doi:10.1016/j.metabol.2009.07.035. Epub 2009 Oct 20

Wang Y, Pessin JE (2013) Mechanisms for fiber-type specificity of skeletal muscle atrophy. Curr Opin Clin Nutr Metab Care 16(3):243–250. doi:10.1097/MCO.0b013e328360272d

Watenpaugh DE, O'Leary DD, Schneider SM, Lee SM, Macias BR, Tanaka K, Hughson RL, Hargens AR (2007) Lower body negative pressure exercise plus brief postexercise lower body negative pressure improve post-bed rest orthostatic tolerance. J Appl Physiol 103(6): 1964–1972

Wells KE, Torelli S, Lu Q, Brown SC, Partridge T, Muntoni F, Wells DJ (2003) Relocalization of neuronal nitric oxide synthase (nNOS) as a marker for complete restoration of the dystrophin associated protein complex in skeletal muscle. Neuromuscul Disord 13:21–31

Witze A (2014) Space-station science ramps up. Nature 510(7504):196–197. doi:10.1038/510196a

Yamasaki M, Shimizu T (2004) Animal experiments using mammals are essential to the study on the mechanisms of phenomena produced in the human body in the space environment. Biol Sci Space 18:102–103

Yang TB, Zhong P, Qu LN, Yuan YH (2003) Space flight and peroxidative damage. Space Med Med Eng (Beijing) 16:455–458

Zehr EP, Stein RB (1999) What functions do reflexes serve during human locomotion? Prog Neurobiol 58:185–205

Zhou L, Zhu DY (2009) Neuronal nitric oxide synthase: structure, subcellular localization, regulation, and clinical implications. Nitric Oxide 20:223–230. doi:10.1016/j.niox.2009.03.001

Chapter 3
Neuromuscular System

Abstract Skeletal muscle adaptations in different environment conditions are particularly sensed at the neuromuscular junction (NMJ), the most distal connection of the neuromuscular system to peripheral target muscles. For example, in all models of muscle disuse atrophy (e.g., bed rest, denervation, limb immobilization, unloading), the NMJ experiences disruptive structural adaptations that are very similar to those observed during exposure to microgravity (μG) in Space. Thus, prolonged disuse and μG both exert adverse effects on the determination of skeletal muscle mass. However, studies on the molecular mechanisms of the normal neuro-muscular system including the NMJ in disuse and in different adaptation conditions are scarce. The NMJ on its own is a highly specialized cell-cell communication contact between a nerve (axon terminal) and a muscle fiber (myofiber) providing the basis for the appropriate peripheral motor control of the neuromuscular system. Altered muscle contraction activity more extensively stimulate NMJ remodeling with synaptic microstructure and size changes. The most relevant changes include decreased synaptic vesicle density and neurotransmitter content, but also axon terminal de-/regeneration with regression and sprouting mechanisms. Previous studies in Space reported microgravity-induced cellular and molecular changes at spaceflown rodent NMJs. We found candidates of the Homer protein family (previously found at central synapses of brain) also expressed at the NMJ subsynaptic microdomain region of human skeletal muscle that were however lost in disuse and upregulated by exercise in bed rest, suggesting activity-driven Homer expression mechanisms and functions with unique roles in normal NMJ function and adaptation.

Keywords Skeletal muscle innervation • Neuromuscular junction • Homer proteins • Disuse • Exercise

3.1 Introduction

Skeletal muscles, and in particular those of the lower limb or antigravity postural muscles (Fig. 2.1), undergo several adaptive processes during altered muscle activity or spaceflight missions which are mostly characterized by structural and metabolic changes that are consequently largely responsible for the altered skeletal

© The Author(s) 2015
D. Blottner, M. Salanova, *The NeuroMuscular System: From Earth to Space Life Science*, SpringerBriefs in Space Life Sciences, DOI 10.1007/978-3-319-12298-4_3

muscle structure and function. The μG-induced muscle atrophy is well known during spaceflight programs (in a time-dependent manner), as well as the related reduction in motor performance control in most of the crew members which are responsible for the decrement in skeletal muscle strength and in resistance to fatigue that severely affected both "inflight" and "extravehicular" crew members' activity. Moreover, morphological alterations of the neuromuscular junction (NMJ) in response to reduced neuromuscular activity and the role of exercise in maintaining the integrity have been extensively investigated since the beginning of the human space program (Wilson and Deschenes 2005; Nishimune et al. 2014).

By using animal experimental models exposed to μG, it was recently shown that different types of muscles experience a greater degree of atrophy than others. In particular, in some of the studies, the high degree of muscle atrophy well correlates with the presence of the slow-type myosin heavy chain suggesting the involvement of a greater proportion of slow-twitch fibers (see Sect. 2.1.5). By contrast, in other reports, atrophy is similarly described in both slow- and fast-twitch fibers, as previously reviewed by Wilson and Deschenes (2005).

One proposed explanation for the relative atrophic response elicited in different muscles is that μG induces a shift in "neuronal recruitment patterns" of motor units (Wilson and Deschenes 2005; Fitts et al. 2001). Thus, we may conclude that nerve endings' remodeling may play an additional and synergistic role in response to altered skeletal muscle loading or altered contraction activity during spaceflight, and the altered neuron-derived specific signals (electrical and/or chemical transmitters) might be, at least in part, responsible for the microgravity-induced effects at the level of single skeletal muscle fiber.

In an attempt to further elucidate the molecular mechanisms which are involved during disuse-induced muscle atrophy, we investigated whether or not scaffold adapter proteins of the Homer family were involved in skeletal muscle plasticity changes. Thus, by confocal immunohistochemistry tools, we explored *soleus* and *vastus lateralis* NMJs of healthy subjects that underwent long-term bed rest muscle disuse without and with two different exercise countermeasures. The novel results of our experiments suggested that components of the Homer family are specifically expressed at the NMJ postsynaptic regions of postural skeletal muscles of healthy human subjects and, surprisingly, they were particularly differently regulated at the protein and mRNA level by muscle activity and inactivity (Salanova et al. 2011).

3.2 Where the Nerve Meets the Muscle: Neuromuscular Synapse

Skeletal muscle contraction is under the tight control of motor nerve activity by means of a unique muscle and nerve common structure named the "neuromuscular synapse" or "neuromuscular junction" (Fig. 3.1).

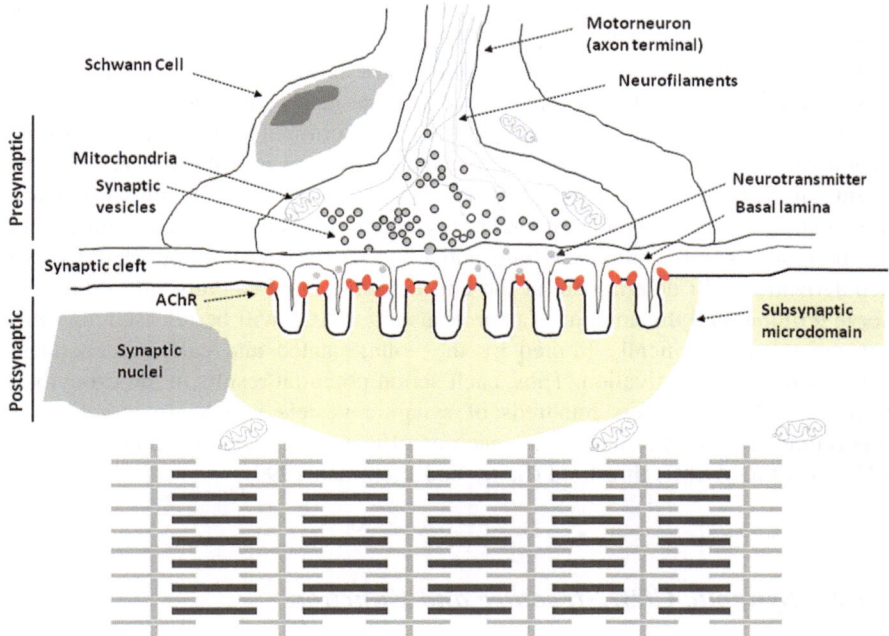

Fig. 3.1 Schematic diagram representing the different compartments of the skeletal muscle NMJ and the contractile apparatus (myofibrils with sarcomeres in series) of the myofiber sarcosol underneath

The NMJ or motor endplate could be functionally and anatomically divided into four distinct regions: a presynaptic, a synaptic (cleft), a postsynaptic region, and a subsynaptic microdomain.

3.2.1   Presynaptic Structure and Function

Every myelinated motor axon reaches its peripheral target muscle where it divides into several unmyelinated terminal fibers each of which directly innervates a single myofiber. Altogether, the terminal fibers of a given single motoneuron and all myofibers that are under its tight control are considered as a single motor unit (Sanes and Lichtman 1999). Basically, the terminal fibers contain both potassium and sodium channels which are important for the control of the duration and amplitude of the membrane action potential. The sodium channels however are absent in proximity of the nerve terminal. The nerve terminal contains synaptic vesicles each of which contain several thousand molecules of the typical neuro-transmitter acetylcholine (ACh). The content of a single presynaptic vesicle is considered to reflect a "quantum" of the neurotransmitter being released upon motor nerve stimulation (Sanes and Lichtman 1999).

Once the action potential reaches the nerve terminal, the signal is further propagated in the cell by the opening of voltage-gated calcium channels which are responsible for a series of events at the nerve terminal which end up with the fusion of synaptic vesicles with the plasma membrane and the release of their quanta of ACh into the synaptic cleft (Fig. 3.1). These late events are particularly regulated by the presence of soluble N-ethylmaleimide sensitive factor attachment receptor or SNARE complex (Hong and Lev 2014). The duration of the nerve depolarization will determinate at the end the amount of calcium influx into the nerve terminal and the amount of neurotransmitter ACh will be released, and this appears to be specifically limited by the voltage-gated and calcium-dependent potassium channel activation. Thus, each action potential results in the exocytosis from several tens to few hundreds of synaptic vesicle which are largely in a supernumerary capacity needed to reach a sufficient number for the postsynaptic ACh receptor (AChR) threshold (Sanes and Lichtman 1999).

3.2.2 Synaptic Cleft Structure and Function

The space between the nerve terminal membrane and the myofiber plasma membrane or sarcolemma is called "synaptic cleft" which is a relatively wide (approx. 50 nm) intermembrane gap as compared to those found in typical central synapses (approx. 5–10 nm) of brain and spinal cord (Fig. 3.1). A peculiar microstructure typical for the NMJ is represented by a synaptic cleft basal lamina, a 3D micro network of loose extracellular matrix molecules and thin filaments that allow for transmitter diffusion but also provide mechanical link (via laminins and fibronectin anchoring proteins) between the pre- and postsynaptic membrane required to withstand mechanical stress in normal muscle tone/tension control, but also by contraction shortenings of the myofibers inside (Nishimune et al. 2004). The synaptic basal lamina is also the location of the acetylcholine esterase (AChE). Once ACh is released from the presynaptic vesicles, the neurotransmitter quickly spreads by diffusion in the synaptic cleft within a few microseconds in order to reach the specific postsynaptic membrane receptors (AChR). The transmitter signal is turned off by AChE enzyme activity present in the synaptic cleft that specifically hydrolyzes the ACh (Aldunate et al. 2004; Gaspersic et al. 1999) neurotransmitter (i.e., acetyl residue plus choline) for instant reuptake by the presynaptic nerve terminal for de novo transmitter synthesis.

3.2.3 Postsynaptic Membrane Structure and Function

The opposite part of the nerve terminal in the close vicinity of the synaptic cleft is called the postsynaptic membrane which actually is a specialized part of the outer muscle fiber membrane or sarcolemma. At this region, the sarcolemma forms several folds and troughs typical of the NMJ (Fig. 3.1). The "entrance" of the little troughs is occupied by the nicotinic acetylcholine receptors (nAchR) at a concentration range approx. of 10.000–20.000 receptors per mm^2 of postsynaptic surface area, at least in a concentration range of about 1:1000 compared to extrasynaptic sarcolemmal regions of a normal myofiber (Fig. 3.1). The strong nAChR density at this region is to ensure for optimal postsynaptic neuromuscular transmission.

Normally, the nAChR consist of pentameric transmembrane subunits which cluster together forming a pore that in absence of ACh molecules is maintained in a closed state in order to ensure uncontrolled flow-through of cations. In the presence of ACh which selectively binds the N-terminal domain of the nicotinergic AChR a1 subunit, exactly two molecules of ACh are required to activate the receptor (Harpsoe et al. 2011; Gotti et al. 2009). Then, the channel pore opens, and sodium moves into the muscle cell and allows for an action potential to propagate forward on the sarcolemmal membrane. Once the action potential reaches the t-tubules (TT), it activates the voltage-gated ion channel dihydropyridine receptor 1 alpha (DHPR1α) by inducing conformational changes that, in turn, modify its physical interaction with the intracellular calcium release channel ryanodine receptor type 1 (RyR1) which is positioned at the sarcoplasmic reticulum terminal cisternae (TC) (Flucher and Franzini-Armstrong 1996) at the interface between the TT (Fig. 2.4). As RyR1 physically dissociates from DHPR1α, it becomes activated and opens providing most of the calcium influx required for muscle contraction (Fig. 2.19). RyR1 exist in the cell as a macromolecular complex whose regulation is under the tight control of different signaling pathways. Altered RyR1 regulation is considered to be a cause of several clinical skeletal muscle myopathies such as central core disease (CCD) or malignant hyperthermia (MH) (Brislin and Theroux 2013; Pietrini et al. 2004).

3.3 Molecular Mechanism of NMJ Adaptation to Disuse and Exercise

There is a large body of evidence suggesting that skeletal muscles, and in particular lower limb and antigravity postural muscles, undergo several adaptive processes during prolonged inactivity or unloading during exposure to microgravity (μG) mostly characterized by structural and metabolic changes, which thereafter might be largely responsible for the altered skeletal muscle structure and function. For instance, the μG-induced muscle atrophy is well known since the beginning of the human space program, as well as the related reduction in motor performance in astronauts, which is the likely one cause for the decrement in skeletal muscle mass and strength and in resistance to fatigue, severely affecting both "inflight" and "extravehicular" crew members' mission duties.

Needless to say that the beneficial effects of exercise for the musculoskeletal quality and neuromuscular control are well accepted today. However, studies on mature NMJs of normal adult skeletal muscle showed continuous remodeling mechanisms even under physiological conditions suggesting a more or less continuum microstructural representation in normal healthy skeletal muscle depending on everyday moderate activities and fitness, throughout human life performances, and even chaperoning healthy aging on Earth that need to be considered for investigations. For instance, extended periods of increased or decreased neuromuscular activity more extensively stimulate remodeling, including changes in the overall NMJ microstructure and size. Recently, the role of exercise in maintaining the integrity of the NMJ at molecular level has been extensively reviewed (Nishimune et al. 2014). Overall, the results suggest that NMJ hypertrophy and the increase in active zone protein levels in exercised young, adult, and aged NMJs are all localized effects within the exercised muscles and motoneurons innervating exercised muscles (Nishimune et al. 2014).

Interestingly, in human skeletal muscle, resistive exercise training upregulates the expression of laminin β2 at mRNA and protein level (Gordon et al. 2012; Hunter et al. 1989) which, in turn, may play a role in the effect of exercise that increases the level of active zone protein Bassoon in aged NMJs (Nishimune et al. 2012). Laminin β2 is an extracellular matrix protein that is secreted by muscles and is concentrated specifically at the synaptic cleft basal lamina of NMJs (Nishimune et al. 2004; Hunter et al. 1989) which binds specifically P/Q- and N-type voltage-dependent calcium channels (VDCCs) (Nishimune et al. 2004). Several in vivo studies have demonstrated that the interaction between laminin β2 and VDCCs is responsible for organizing the NMJ active zones. The number of active zones, in fact, was decreased when the interaction between the VDCCs and laminin β2 was experimentally perturbed in wild-type mice (Nishimune et al. 2004). Consistent with the above reports, P/Q- and N-type VDCC double-knockout mice exhibited specific defects in the number of active zones and docked synaptic vesicles, which were twice as severe as the defects observed in the P/Q- and N-type VDCC single-knockout mice (Nishimune et al. 2004; Chen et al. 2011). Moreover, humans who carry laminin β2 mutations result in active zone loss and denervation in addition to the development of the "Pierson syndrome," an autosomal recessive movement disorder associated with microcoria, and nephrotic syndrome (32, 49). These results suggest that laminin β2 binds to synaptic VDCCs to organize the active zones. Recently, studies showed that NMJ active zones are maintained at a constant density as the NMJ matures but are degraded in aged animals (Chen et al. 2012). On the same line of evidence, the expression level of Bassoon decreases in the NMJs of aged mice and rats (Nishimune et al. 2012; Chen et al. 2012). A lack of Bassoon is known to impair synaptic vesicle trafficking to presynaptic membranes in the central nervous system and sensory neurons (Hallermann et al. 2010; Mukherjee et al. 2010; Frank et al. 2010). Furthermore, a lack of Bassoon decreases

VDCC Ca^{2+} influx and weakens synaptic transmission, because of the direct interaction between VDCCs and Bassoon that enhances the P/Q-type VDCC function (Nishimune et al. 2012). This modification to VDCCs by Bassoon is similar to the effect of another active zone protein such as RIM1 on VDCCs (Kiyonaka et al. 2007; Uriu et al. 2010). If confirmed, these findings suggest that active zone protein loss may be a part of the molecular mechanism that causes the deterioration of aged NMJs. Active zone deterioration in aged NMJs is ameliorated by muscle exercise. Two months of isometric force training rescued the loss of Bassoon in aged NMJs in the genioglossus muscle of 2-year-old rats (Nishimune et al. 2012). However, exercise training did not alter NMJ size, which suggests that the increase in the Bassoon immunohistochemistry signal in aged NMJs reflects an increase in the protein quantity in each nerve terminal (Nishimune et al. 2012). The mean intensity of the Bassoon immunohistochemistry signal in the NMJs of exercised aged rats is similar to the mean intensity observed in young adult rats. Importantly, the improvement in Bassoon protein level observed in exercised and aged NMJs is consistent with improvements observed using electrophysiology in NMJ function after endurance training in aged mice (Fahim 1997). In summary, exercise-induced upregulation of laminin β2 may play a role in preservation of active zones in aged NMJs, which more likely exerts a positive effect on NMJ synaptic transmission.

3.4 Neuromuscular System Adaptation to Spaceflight

Since the beginning of the "human space program," most of the space research activity has dealt with the effect of μG on bone and skeletal muscle adaptation. For skeletal muscle in particular, most of the studies were made primarily at the physiological and behavioral levels, while only few studies so far investigated molecular and biochemical NMJ adaptation to μG exposure. Nevertheless, these studies however reported that μG during spaceflight induced disturbance in neuromuscular transmission together with a significant decrement in muscle strength (Deschenes et al. 2005).

Data recently available on the histologic and ultrastructural level generally reported that one of the most disruptive NMJ adaptations to μG is axonal disintegration and substantial decrease in synaptic vesicle density and neurotransmitter content, although small quantitative differences were often present between different types of muscle (Baranski and Marciniak 1979; Deschenes et al. 2005). Changes in NMJs of antigravity postural muscles such as *soleus* (SOL) or adductor longus, as reported, were always more pronounced than other muscles such as extensor digitorum longus (EDL) or gastrocnemius (GAS) (Riley et al. 1990). Interestingly, by using in vivo animal experiments, it was proposed that resistance training in rat SOL is able to remodel structure and size of the NMJ by increasing nerve terminal branching during muscle unloading (Deschenes et al. 2000, 2006). Moreover, neuromuscular adaptations in spaceflown rats were further investigated (Deschenes et al. 2005), which confirmed that these changes are present in particular in postural muscles or antigravity muscles such as the soleus (SOL) and were clearly manifested at the neuromuscular synapse. Thus, the general emerging picture is that under exposure to µG, NMJ experiences disruptive structural adaptations similar to those observed in several models of muscle disuse atrophy on ground such as bed rest inactivity, limb immobilization, denervation, or animal model of hind-limb unloading. Such muscle adaptations might be further accompanied at the level of NMJ by molecular modifications of specific cell signaling pathways and scaffold adapter proteins particularly involved in the synaptic transmission.

3.5 Role of Scaffold Adaptor Homer Proteins in Skeletal Muscle

So far there are several studies suggesting the importance of Homer protein family in skeletal muscle physiology (Stiber et al. 2005). For instance, specific isoforms of the Homer family have been proposed to directly modulate the intracellular calcium release channel ryanodine receptor 1 (RyR1) activity (Feng et al. 2008). Moreover, binding of Homer modulates the activity of various Ca^{2+} channels, whereas the formation of multimers allows cross-talk between different surface membrane receptors and Ca^{2+} channels in the membrane of intracellular compartments.

Primarily Homer proteins were originally isolated from central nervous system (CNS) neurons as one of the several proteins whose expression is upregulated upon synaptic activity induced by seizure or during induction of long-term potentiation (Shiraishi-Yamaguchi and Furuichi 2007). Homers can bind proline-rich regions present on target proteins through their EVH1 domains and cross-link them through their coiled-coil domains discussed in a recent review (Salanova et al. 2013). The Homer short forms, including Homer1a, the IEG, that directly modify synaptic plasticity (Shiraishi-Yamaguchi and Furuichi 2007) cannot form multimers, but they can still bind to target proteins through their EVH1 domain thereby modifying protein function. Thus, Homer1a disrupts signaling complexes, and therefore, this protein isoform has been proposed to function by acting as a negative regulator of the long forms of Homers (Xiao et al. 1998; Roche et al. 1999).

The EVH1 domain selects a specific consensus binding site within different target proteins that represents a short proline-rich amino acid sequence, PPXXΦ (where X is any amino acid and Φ is any aromatic amino acid), with charged residues on either side of the sequence (Salanova et al. 2013). Database screening analysis showed that there is a long list of potential candidates of Homers interacting proteins, which contained a poly-proline-rich consensus sequence. Some functional proteins were already shown to interact with Homers, e.g., the G-protein-coupled metabotropic glutamate receptor (mGluR) and several Ca^{2+}-signaling proteins, including Shank, PLCb, IP3Rs, TRPC channels, RyR, several

L-type Ca^{2+} channel isoforms and the nuclear factor of activated T-cells (NFAT) (Huang et al. 2008).

Homers regulate the myogenic program since the transient expression of specific Homer protein isoforms precedes myoblast fusion and thus myotube formation/ skeletal muscle differentiation (Stiber et al. 2005). Further evidence supporting this hypothesis comes from findings indicating that Homer 2 mRNA and protein were found to be transiently expressed at high levels in whole embryo at the stage E14.5, particularly in the brain and hind limbs (Stiber et al. 2005). Further biochemical evidence for an effect of Homer 2b isoform on muscle differentiation was ascertained by examining the expression of RyR1 and of the myoglobin promoter, two well-established muscle-specific differentiation markers. Moreover, C_2C_{12} cells (mouse cell line) expressing Homer 2b augmented basal NFAT transactivation and myoglobin promoter in the order of 2.5-fold greater than of empty vector transfected control cells. Interestingly, co-expression of both activated calcineurin and Homer proteins resulted in a synergistic activation (16-fold) of the myoglobin promoter (Stiber et al. 2005). A proposed mechanism was that Homer might play a role as a molecular switch mediating both IP3R and RyR signaling to promote myotube differentiation program through the NFAT signaling pathway. Thus, Homer proteins potentially represent the first example of a family of scaffolding proteins, which are involved in modulating Ca^{2+}-dependent gene expression during muscle differentiation (Stiber et al. 2005).

An additional role of Homer protein family concerning the regulation of excitation-contraction (E-C) coupling was also proposed and suggested by the presence of a Homer binding motif in all RyRs and voltage-dependent L-type Ca^{2+} channels. This hypothesis was further confirmed by specific Ca^{2+} release measurement experiments in permeabilized isolated skeletal muscle fibers as well as by the analysis of the open probability (*Po*) of the RyRs channels inserted into lipid bilayers, which showed that Homers are able to activate both types of ryanodine receptor Ca^{2+} channels RyR1 and RyR2. Other studies reported that co-expression of Homer1a with RyR2 and $Ca_v1,2$ enhances, whereas co-expression of Homer1b with RyR2 and $Ca_v1,2$ reduces, the efficiency of Ca^{2+}-induce Ca^{2+} release (CICR) in an in vitro cell model (Huang et al. 2007).

3.5.1 Mice with a Specific Deletion of Homer1 Develop Skeletal Muscle Myopathy

Homer1 interacts with several members of the transient receptor potential classic (TRPC) family acting as a scaffolding protein and modulating the formation of a large protein complex which involves TRPC1 and IP3R channels. Recently, Stiber et al. (2008) reported that Homer1$^{-/-}$ mice exhibit TRPC1 abnormalities such as altered ion channel activity, which leads to a pronounced skeletal muscle myopathy, as revealed by myofiber cross-sectional area analysis. Furthermore, Homer1$^{-/-}$

(KO) mice exhibited a decrease in skeletal muscle force contractility (Stiber et al. 2008). It was speculated that Homers might be localized at the NMJ and lack of Homer1 in KO mice appeared to mimic mechanisms of denervation (Stiber et al. 2008). However, the first evidence of Homers at NMJs in human and rodents was found and described more recently by our group (Salanova et al. 2011).

3.5.2 Skeletal Muscle Regeneration Characterized by Transition of Homer-specific Isoforms

Transition in Homer isoforms under resting conditions and under short- and long-term muscle adaptation has been previously investigated (Bortoloso et al. 2006). Interestingly, a switch in Homer-specific isoforms was proposed to be associated with muscle differentiation and regeneration mechanisms (Bortoloso et al. 2006). For instance, constitutively expressed Homer1a is transiently upregulated during myofiber regeneration and myotube formation. By contrast, Homer1b/c/d, −2b expression positively correlated with changes in muscle mass (Bortoloso et al. 2006). Therefore, a pivotal role of Homers in skeletal muscle plasticity and adaptation appears very likely.

3.5.3 Regulation of Homer Expression in Denervated Myofibers

By using a rat sciatic denervation model, the expression of Homer1b/c and Homer2a/b was investigated in detail (Bortoloso et al. 2013). Only Homer2 was shown to be downregulated at mRNA and protein level in atrophic muscle after 7 days of permanent denervation (in vivo unilateral sciatic nerve transection), suggesting a neuronal control of Homer expression in skeletal muscle. Interestingly, reconstitution of Homer2 by in vivo transfection experiments in denervated soleus partially rescued the atrophic phenotype (Bortoloso et al. 2013).

3.5.4 Hind-Limb Unloading (HU) Decrease Homer Expression

The expression of specific Homer isoforms has been investigated in rat soleus following hind-limb unloading (Bortoloso et al. 2013). Seven days of HU were sufficient to decrease Homer2 protein expression, which in turn affects the ubiquitination and thus, indirectly, the transcriptional activity of atrophy-related genes such as MuRF1, atrogin, and myogenin (Bortoloso et al. 2013). In fact, downregulation of Homer2 appears to be an early event in muscle atrophy.

Additional studies from our group using the HU model showed that loss of Homer proteins at the NMJ of rat soleus after 21 days of HU was clearly attenuated by an "active mode motion" (*Salanova et al., manuscript in preparation*); an exercise protocol which is characterized by spontaneous muscle contractions in HU rats entrained by electric impulse stimulation of the hind foot paw (Sun et al. 2013). Whereas, "passive mode motions" (an exercise protocol triggered by mechanically-induced hind limb lifting passively (Sun et al. 2013)) did not show similar effects (*Salanova et al., manuscript in preparation*).

3.5.5 Regulation of Myogenic Differentiation Program by Homer

Recently, it has been proposed that Homer-specific isoforms are regulators of myotube formation and, thus, inducers of the myogenic differentiation program. Additional evidence supporting this hypothesis stems from results of Homers that were found to be associated with several regulatory proteins, including Drebrin, a F-actin side-binding protein, and the small GTPase Cdc42 (Mancini et al. 2011). Moreover, Homer2b was found to be in the calcineurin/nuclear factor of activated T cells (NFAT) pathway and to enhance myoblast differentiation and myotube formation (Stiber et al. 2005).

3.5.6 Homer Localize in Skeletal Muscle-Specific Subcellular Compartments

Anti-pan Homer antibodies revealed two different immunofluorescence patterns. In longitudinal sections, a cross-striated sarcomeric labeling at the Z-line/costamere level was present, corresponding to the longitudinal sarcoplasmic reticulum cisternae (LSR) (Salanova et al. 2002). A second, moderate but discrete Homer immunolabeling punctuate pattern was observed at the NMJ corresponding to the NMJ subsynaptic microdomain. As shown in Fig. 3.2, α-bungarotoxin labeling of AChRs localized adjacent to the Homer antigens (Salanova et al. 2011).

3.5.7 Role of Homers in Skeletal Muscle Atrophy

The signaling pathways responsible for muscle mass control/atrophy during prolonged disuse are not yet fully understood. Searching for new molecular players responsible for disuse-induced muscle atrophy, we recently proposed that proteins

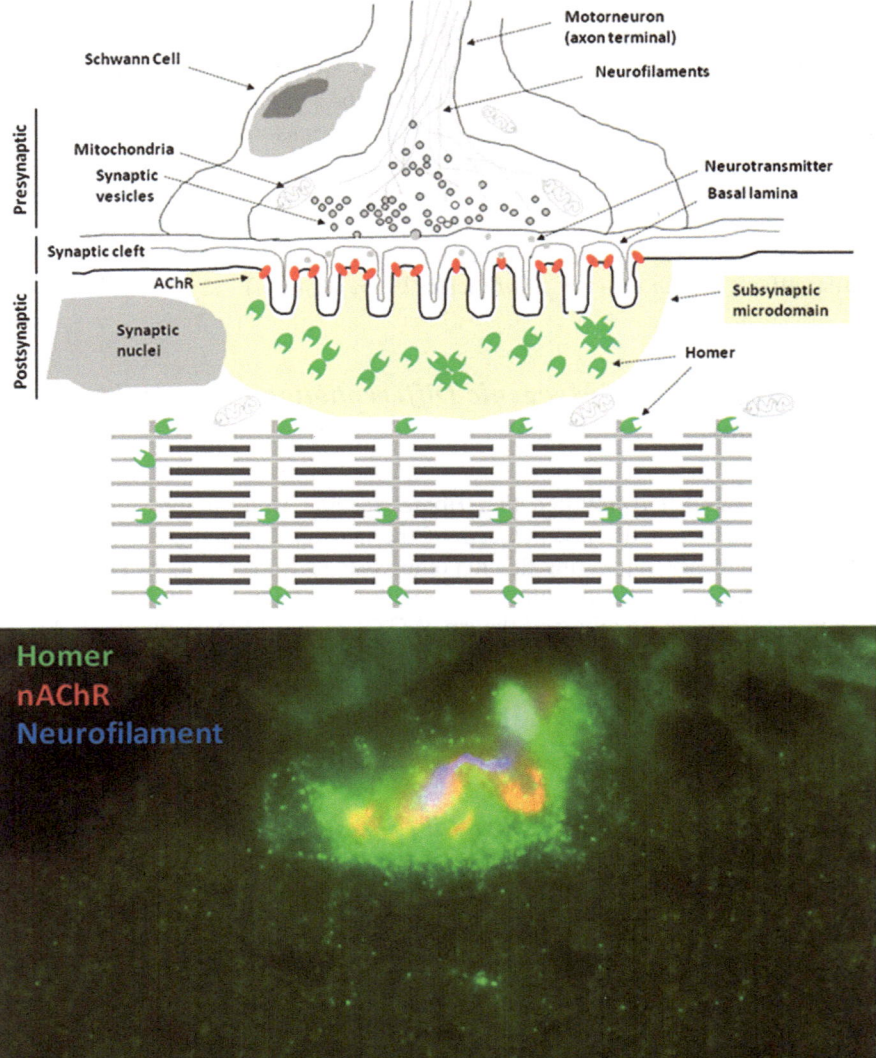

Fig. 3.2 Homer protein is concentrated at the NMJ. *Upper panel*, diagram representing Homer protein subcellular localization in human skeletal muscle. Homer proteins (*green dots*) are expressed both at the NMJ subsynaptic microdomain and at the Z-lines (costamere microdomain) of the contractile apparatus (Salanova et al. 2013). *Lower panel*, immunofluorescence image of human soleus myofiber with NMJ (red, nAChRs) immunostained with anti-Homer and anti-neurofilament (*blue*) antibodies

of the Homer family might play a pivotal role in neuromuscular synaptic signaling (Salanova et al. 2011). Based on our observation that Homers are particularly concentrated at the postsynaptic membrane of human and rodent neuromuscular junctions (NMJs), we suggested that they may play a role in the calcineurin-NFATc1 signaling pathway (Salanova et al. 2011).

By using a mouse KO (gene knockout) model, it has been shown that the ablation of the Homer1 gene leads to pronounced skeletal muscle myopathy associated with the increased transient receptor potential C1 (TRPC1) activity and, thus, calcium influx (Stiber et al. 2008). Furthermore, inducing muscle disuse atrophy by using the hind-limb unloading (HU) model, recently it was observed that 3 weeks of HU resulted in a significant decrease of Homer proteins (Bortoloso et al. 2013) that was also observed at the NMJ of atrophic soleus myofibers (Salanova et al., unpublished results). Interestingly, results obtained in animal models well correlated with the decrement of Homer mRNA and proteins at the NMJs of human subjects who underwent 60 days of 6° head down tilt (HDT) bed rest muscle disuse (Salanova et al. 2011). Taken together, the data indicate that synaptic expression of Homers is regulated by motor nerve activity and that the Homer protein family plays an important yet underestimated role in the pathophysiology of normal skeletal muscle function and in disuse atrophy likely affecting normal neuromuscular system control.

3.6 Homer Proteins in Bed Rest: Elucidating Signaling Pathways

As already discussed, skeletal muscle contractions are based on signal transmission from a motor nerve to its skeletal muscle at special nerve-muscle contacts known as the neuromuscular junction (NMJ) or neuromuscular synapse. As predicted, a direct comparison between resistive exercise (RE) and resistive vibration exercise (RVE) training protocols performed by two exercise groups on the same Galileo trainer in bed rest (2nd ESA Berlin Bedrest Study, BBR-2) showed that RVE particularly affected neuromuscular control mechanisms driven by nerve-muscle interactions (Salanova et al. 2011). Molecular analysis combined with high-resolution confocal laser analysis of NMJ structures in muscle biopsy material from the BBR-2 study showed that RVE and to lesser extent also RE are preferentially able to upregulate Homer2a, a protein localized at the neuromuscular junction subsynaptic microdomain of both SOL and VL muscle (Fig. 3.3).

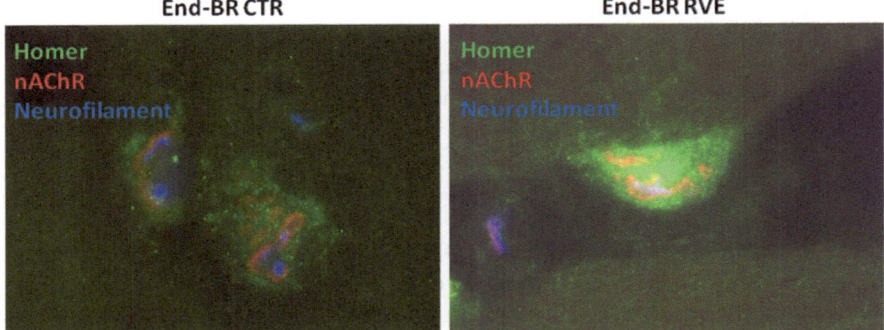

Fig. 3.3 Homer protein is regulated at the NMJ by muscle activity. *Left panel*, immunofluorescence image of SOL of bed rest CTR group with NMJ (*red*, nAChRs) immunostained with anti-Homer (*green*) and anti-neurofilament (*blue*) antibodies. *Right panel*, immunofluorescence image of SOL of bed rest RVE group with NMJ (*red*, nAChRs) immunostained with anti-Homer (*green*) and anti-neurofilament (*blue*, neurite marker) antibodies. Increased Homer protein (*green dots*) fluorescence spot density and intensity are clearly visible at the NMJ subsynaptic microdomain in a RVE trained bed rest subject (Salanova et al. 2011, 2013)

3.6.1 Resistive Vibration Exercise Increased NFATc1 Proteins at the NMJ Which Fully Correlates with Homer2 Changes

Homer2 proteins may regulate the calcineurin (CaN)-dependent NFAT pathway, an established control mechanism for the specification and maintenance of the slow-twitch muscle phenotype. Given this assumption, we determined whether NFATc1 proteins co-localized with Homer at the NMJ of human skeletal muscles. Our results showed that NFATc1 immunoreactivity is detectable at the NMJ subsynaptic compartment as judged by α-bungarotoxin co-labeling of AChR, the immunofluorescence pattern being similar to that obtained with anti-Homer antibodies in serial sections (not shown). Comparative and quantitative fluorescence intensity analysis between end-bed rest CTR and RVE samples clearly showed that NFATc1 expression patterns paralleled those of Homer at the very same NMJ structures (Fig. 3.4).

3.6.2 Homer Scaffold Might Be Required for NFATc1 Recruitment at the Skeletal Muscle NMJ

As suggested by immunohistochemical colocalization studies, Homer proteins might directly interacted with NFATc1 at the NMJ. This was further supported by biochemical affinity chromatography experiments with a fragment of the Homer EVH1 (Ena/VASP Homology 1) domain (Salanova et al. 2011). Because the first

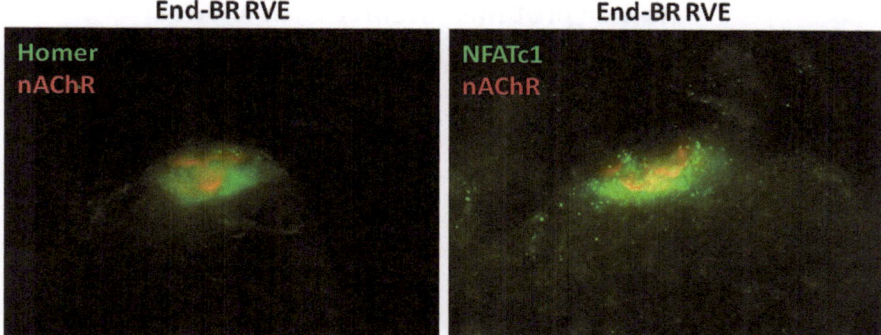

Fig. 3.4 Homer and NFATc1 proteins (specific signaling molecules) are both increased at the NMJ of RVE trained bed rest subjects. *Left panel*, immunofluorescence image of SOL of RVE end-bed rest subject with NMJ (*Red*, nAChRs) immunostained with anti-Homer (*green*) antibodies. *Right panel*, immunofluorescence image of SOL of RVE end-bed rest subject with NMJ (*red*, nAChRs) immunostained with anti-NFATc1 (*green*) antibodies

80 amino acids in the amino-terminal region of the Homer EVH1 domain (N-80) were sufficient to interact with several NFATc isoforms in T cells (Huang et al. 2008), we used two different GST-Homer constructs, Homer 120 (N-terminal 120aa) and Homer 131 (N-terminal 131aa) which both contain the fist N-80 fragment to study the interaction of NFATc1 with Homer in skeletal. The GST alone was additionally used as negative control (empty vector). GST and GST-Homer fusion proteins were then purified on a glutathione column (Fig. 3.5), immobilized to a CNBr-Sepharose, and incubated with muscle homogenates of rat gastrocnemius. GST and GST-Homer protein eluate were investigated for the presence of NFATc1. As shown by our experiment results, anti-NFATc1 antibodies produced the same immunostaining pattern obtained with anti-Homer antibodies at the NMJ (Fig. 3.5) and NFATc1 immunoreactive bands were identified by Western blot analysis in eluate of the GST-Homer N-120 or N-131 column but not in the eluate of the GST column suggesting a specific molecular interaction between Homer and NFATc1 in skeletal muscle (Fig. 3.6).

3.6.3   Resistive Vibration Exercise Significantly Increased NFATc1 Positive Myonuclei

Basically, the nuclear factor of activated T cells (NFAT) is activated by cell-surface receptors coupled to Ca^{2+} mobilization. Increased cytosolic Ca^{2+} concentration results in the activation of a family of Ca^{2+}-dependent calmodulin enzymes including calcineurin. A protein phosphatase which dephosphorylates multiple phosphoserines on the NFAT leads to its nuclear translocation (Luo et al. 1996) and thus activation during electrical stimulation of skeletal muscles with a slow-

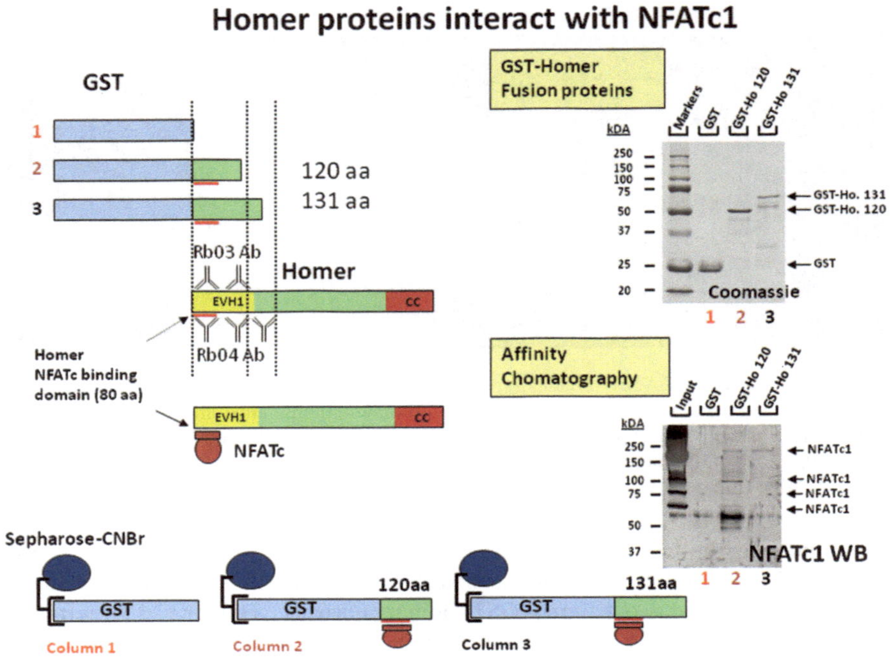

Fig. 3.5 Homer in skeletal muscle is part of the NFAT signaling pathway. *Upper left panel.* Diagram depicting GST and GST-Homer constructs. *Lower left panel.* Homer N-terminal NFATc1 binding site. *Upper right panel.* SDS-PAGE Coomassie brilliant blue staining of purified glutathione S-transferase (GST) and GST-Homer fusion proteins. *Lower right panel.* Western blot analysis with anti-NFATc1 antibodies of GST and GST-Homer eluate of rat muscle homogenates. Note that several NFATc1 immunoreactive bands were pulled down by GST-Homer fusion protein but not by the GST only (empty vector) that was used as negative control (Modified from Salanova et al. 2011)

type firing pattern (Tothova et al. 2006). As predicted, the NFATc1 subcellular distribution in myonuclei of slow-twitch muscle fibers was significantly decreased after bed rest only, while it was increased by vibration exercise in skeletal muscle SOL and VL of BBR2 subjects (Fig. 3.7), suggesting an additional novel role of vibration exercise by both the maintenance of the slow-type myogenic program and preventing myofiber-type shift toward the fast twitch or type II (Salanova et al. 2011). Thus, prolonged bed rest, i.e., chronic muscle disuse, significantly decreased the NFATc1 transcriptional activity in slow-twitch fibers, while vibration exercise combined with resistive exercise (RVE) effectively counteracted such changes. These findings well correlated with our morphological and morphometric analysis (myofiber-type composition and cross-sectional area) of the very same samples (Salanova et al. 2014). Thus, taken together, the results of our experiments suggest that vibration exercise combined with resistive exercise is a reliable countermeasure protocol against neuromuscular impairments in different unloading conditions of skeletal muscle structure and function.

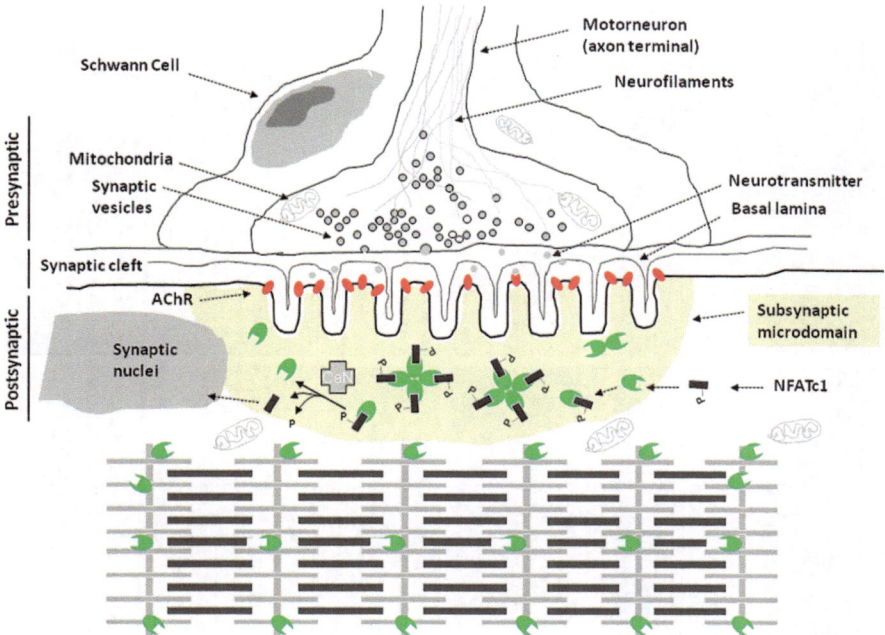

Fig. 3.6 Schematic diagram suggesting hypothetical role of the scaffold adapter Homer protein at the NMJ subsynaptic microdomain. Homer might be part of a macromolecular complex which provides a molecular mechanism responsible for the NFATc1 subcellular localization at the NMJ subsynaptic microdomain

3.7 Vibration Exercise as a Reliable Tool to Stimulate the Myogenic Differentiation Program in C2C12 In Vitro Cell Model

In an attempt to further investigate the biological effects of vibration exercise on skeletal muscle differentiation and maturation, Wang et al. (2010) used cultured C2C12 undifferentiated myoblasts. The major aim was to investigate whether or not vertical vibration was able to stimulate extracellular matrix (ECM) formation which is essential for the myoblast proliferation, migration, and differentiation followed by fusion or myotube formation which could be used as index of myogenesis determination. As reported, the authors concluded that vertical vibration significantly enhances the expression level of several ECM proteins and with them several myogenic regulatory factors such as MyoD, Myf-5, myogenin, and MRF4 (Wang et al. 2010). Interestingly, vertical vibration increased the expression of decorin, a protein which is involved in the interaction with several other proteins including the type I collagen fibrils thus playing a central role during matrix assembly and not less important during muscle hypertrophy. Decorin is secreted from myotubes during exercise and specifically interacts with and inhibits

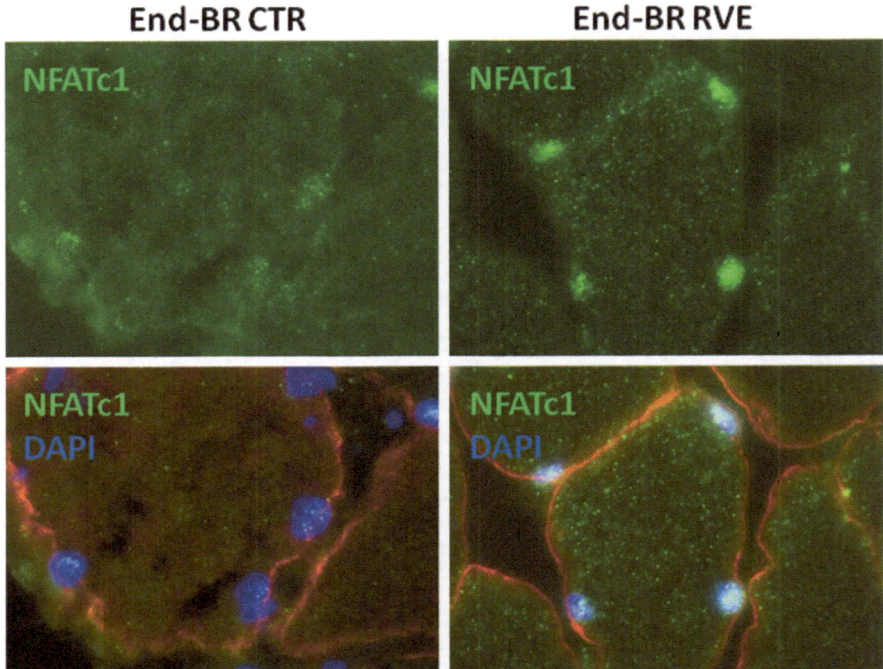

Fig. 3.7 Cytosolic-myonuclear translocation of NFATc1 increased by RVE during bed rest. *Upper left panel.* SOL cryosection of end-bed rest CTR subject immunostained with anti-NFATc1 antibodies. *Lower left panel.* The same cryosection with counterstained myonuclei (DAPI, *blue*). *Upper right panel.* SOL cryosection of end-bed rest RVE subject immunostained with anti-NFATc1 antibodies. *Lower right panel.* The same cryosection with counterstained myonuclei (DAPI, *blue*)

myostatin to promote muscle hypertrophy (Kanzleiter et al. 2014). The neuro-reflexive stimulation of human skeletal muscle by 1a afferent transmission and the H-reflex elicited by vibration exercise has been confirmed in parabolic flight (20 s of transient microgravity exposure), suggesting that the stretch reflexes may not be compromised by microgravity (Kramer et al. 2013). Further studies are planned with crew members using vibration exercise as a potentially effective pre-, in-, and postflight countermeasure co-protocol in combination with currently onboard used countermeasure protocols.

3.8   The NMJ in Spaceflight (Bion-M1 Mission)

In an attempt to further understand the effects of µG on skeletal muscle structure, we have investigated the expression and localization of Homer signaling proteins at the NMJ of skeletal muscle of mice exposed for 30 days to near-orbit spaceflight µG. Preliminary results indicate that the immunodetectable levels of Homer antigens on flown mice muscle, when compared to ground control animals, were significantly decreased, both in skeletal muscle fibers (Z-disk/LSR/costamere Homer protein fraction) and at the NMJ level (Fig. 3.8). These results suggest that µG, either directly or indirectly, affect Homer protein expression and subcellular localization in mouse skeletal muscle and probably affect many other signaling pathways downstream of Homers.

Several reports suggested that an increase in muscle force and power following a given period of vibration exercise are more or less similar to those observed following resistive exercise training (Bosco et al. 1999a, b; Delecluse et al. 2003, 2005). These results well correlated with the study previously reported by Cardinale and Bosco (2003) which reported that both types of exercise, resistive exercise (RE) and resistive vibration exercise (RVE) training, share the neuromuscular system as one major target. However, while during RE training an extra load is required for counteracting the effects of gravity (resistance), resistance load by RVE is overcome by adjusting vibration frequency and amplitude both of which affect acceleration and force exerted on the body. Nevertheless, changes in the level of acceleration may correspond to changes at the level of neuromuscular

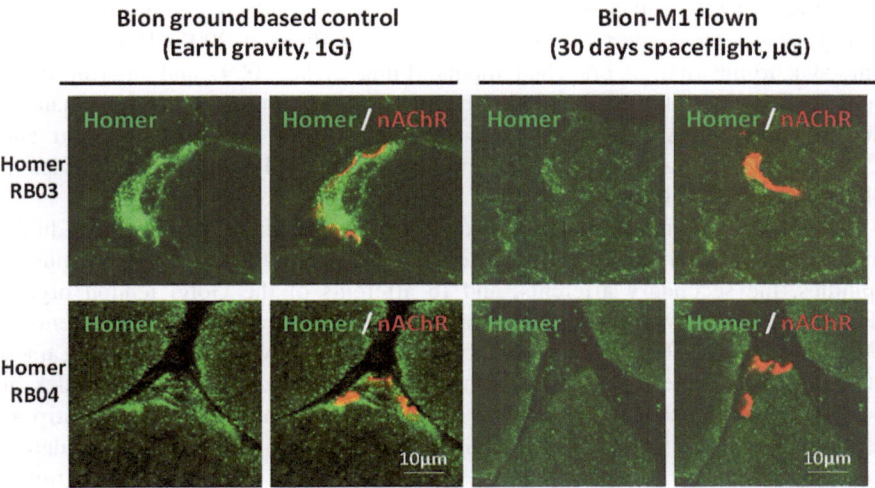

Fig. 3.8 NMJs (*red*) of adult muscle soleus mice housed on the Bion-M1 module on ground (1G) and on Bion-M1 module on orbit (µG) for 30 days. Immunostaining was performed with two different affinity purified pan-Homer antibodies (rabbit Rb03 and Rb04). (Salanova et al., preliminary results)

adaptations which, in turn, may affect levels of, for example, neurogenic and myogenic growth factors. Thus, any changes normally observed during resistive and/or vibration exercise trainings were so far considered as neuronal adaptations (Aagaard et al. 2002; Gabriel et al. 2006; Staron et al. 1994), suggesting that in both training conditions, neurotransmission and signaling at NMJ structures seem to play unique and pivotal roles yet to be further experimentally investigated.

The positive outcome of muscle vibration exercise on disused skeletal muscle in bed rest and the changes found in Homer-NFATc1 myocellular signaling induced by RVE may be at least partly due to mechanical vibration signals targeted directly to skeletal myofibers (e.g., stretching and shortening mechanisms of signal transmission) or may be due to neuronal stimulation via efferent and afferent peripheral nervous system stimulation (e.g., skin, fascia or muscle proprioceptors, central neural pathways, brainstem, and spinal reflexes) that are highly complex and need to be further investigated in more detail.

One great challenge might be to fully understand the underlying mechanisms of vibration mechanosignals in biological tissues and their cellular signaling pathways in skeletal muscle and neuromuscular structures. For example, how are vibrations mechanosignals converted into biochemical cell modifications at the level of the NMJ? A probable explanation is that the numbers of impulses per second or "motor unit firing frequency" that is received by each myofiber within the same motor unit from its own α-motoneuron are dramatically increased by vibration signals. Consistent with this hypothesis is the observation that vibration alone was able to increase force, EMG, and firing frequency rate during fatiguing isometric maximal voluntary contractions (Bongiovanni and Hagbarth 1990). Moreover, muscle spindles are the sensing/responsive elements to vibration, especially the Ia-afferents (Burke et al. 1976; Roll et al. 1989) which may prevent a decline in muscle spindle activity in order to support the motoneuron pool (Griffin et al. 2001). But also motor unit synchronization in branched inputs (Fling et al. 2009) and intermuscular coordination that allowed interaction between muscle groups might be enhanced during vibration. Abercromby et al. (2007) supported by others (Mischi and Cardinale 2009) suggested that both may play additional and synergistic roles in physiological vibration sensing.

Additional evidence further supported the idea that direct application of high-frequency vibration on muscle and tendon activates Ia-afferent of the muscle spindles, the secondary afferents, and Ib afferents of the Golgi tendon organs (Roll et al. 1989) which are all capable of generating evoked cortical potentials (Munte et al. 1996) resulting in corticospinal pathway excitation (Carson et al. 2004). Conversely, more contained vibration frequency in the order of below 20 Hz was capable to activate human Ia and II afferent firing in vitro as reported by Burke et al. (1976) which stimulates the slow-type or oxidative myogenic program. All authors so far speculated that the H-reflex pathway might be the mechanism which converts the vibration stimulus into descending α-motoneuron impulses (Fornari and Kohn 2008; Person and Kozhina 1992). However, some discrepancies still exist in the literature that requires further investigations.

3.9 Risk Factors and Possible Constraints Using Whole Body Vibration Stimulation

From several bed rest studies, we learned that vibration training is a short duration and reliable countermeasure protocol to adequately preserve skeletal muscle structure and function. We know that in response to vibration stimulus, muscle contracts which is mostly due to a "tonic vibration reflex" discovered since 1960s (Hagbarth and Eklund 1966) measured by electromyography (EMG). However, by using vibration as physical intervention to prevent muscle and bone atrophy, several studies pointed toward possible limitations for uncontrolled applications on the human body due to several possible side effects in particular when the system appears to be overstimulated, thus taking serious care of the individual training conditions of potential users. Indeed, it has been experienced that above a certain level of amplitude and frequency (e.g., above 35 Hz) but also of extended time duration or excessive training, bouts of some users might lead to various damaging effects of the human body that go from headache to internal bleeding (Mester et al. 2006). In particular, the cardiovascular system appears to be very responsive to vibration signals that might modify or even increase total peripheral resistance of blood artery and capillaries (Mester et al. 2006). Several investigations recently introduced a term of "resonance frequency" which is the frequency where skeletal muscle reaches maximal transmissibility by vibration, for example, in the order of very low 5–10 Hz (Mester et al. 2006), or even higher frequencies above 20–25 Hz that should be avoided during vibration muscle training. Therefore, there are some constraints in the potential application of vibration stimulation for subjects suffering from coronary disease, hypertension, or osteoporosis or those participating in various rehabilitation programs. Thus, particular care must be given to individually establish appropriate and safety training protocols for potential users in routine fitness programs or in various clinical settings. Due to the obvious health and fitness conditions in crew members, similar constrains and risk factors of vibration muscle exercise may not be a problem for inflight countermeasure applications in future missions to Space.

Conclusion and Perspectives
Currently, a huge body of experimental evidence is found in the relevant literature that emphasizes some of the positive effects of vibration on the preservation of the skeletal muscle structure and function in different environment conditions. For instance, vibration exercise studies on normal human skeletal muscle from the last decade have shown benefits in sport exercise (Mester et al. 2006; Cochrane 2011; Ritzmann et al. 2014), in sarcopenia in the elderly (Bogaerts et al. 2007), but also in various clinical settings such as for patients with neurodegenerative diseases (Sitja Rabert et al. 2012), and in

(continued)

Duchenne muscular dystrophy (Soderpalm et al. 2013). These and many other studies on vibration exercise on humans and also in different animal models of muscle disuse (mostly mice and rats) currently found in the literature are good examples of how findings from fundamental basic research related to Space Life Sciences may be translated to practical applications of completely new and most efficient and probably also more compliant exercise protocols to mitigate or even prevent disuse atrophy of apparently healthy populations, which is a great challenge of our modern sedentary society on Earth, but also a great challenge for the crew members' performance control during their prolonged mission duties on the ISS, during extravehicular activities on orbit, or in future Moon or Mars missions, and their recovery after coming home to Earth.

References

Aagaard P, Simonsen EB, Andersen JL, Magnusson P, Dyhre-Poulsen P (2002) Increased rate of force development and neural drive of human skeletal muscle following resistance training. J Appl Physiol (1985) 93:1318–1326

Abercromby AF, Amonette WE, Layne CS, McFarlin BK, Hinman MR, Paloski WH (2007) Variation in neuromuscular responses during acute whole-body vibration exercise. Med Sci Sports Exerc 39:1642–1650

Aldunate R, Casar JC, Brandan E, Inestrosa NC (2004) Structural and functional organization of synaptic acetylcholinesterase. Brain Res Brain Res Rev 47:96–104

Baranski S, Marciniak M (1979) Stereological ultrastructural analysis of the axonal endings in the neuromuscular junction of rats after a flight on Biosputnik 782. Aviat Space Environ Med 50: 14–17

Bogaerts A, Delecluse C, Claessens AL, Coudyzer W, Boonen S, Verschueren SM (2007) Impact of whole-body vibration training versus fitness training on muscle strength and muscle mass in older men: a 1-year randomized controlled trial. J Gerontol A Biol Sci Med Sci 62:630–635

Bongiovanni LG, Hagbarth KE (1990) Tonic vibration reflexes elicited during fatigue from maximal voluntary contractions in man. J Physiol 423:1–14

Bortoloso E, Pilati N, Megighian A, Tibaldo E, Sandona D, Volpe P (2006) Transition of Homer isoforms during skeletal muscle regeneration. Am J Physiol Cell Physiol 290:C711–C718

Bortoloso E, Megighian A, Furlan S, Gorza L, Volpe P (2013) Homer 2 antagonizes protein degradation in slow-twitch skeletal muscles. Am J Physiol Cell Physiol 304:C68–C77

Bosco C, Cardinale M, Tsarpela O (1999a) Influence of vibration on mechanical power and electromyogram activity in human arm flexor muscles. Eur J Appl Physiol Occup Physiol 79:306–311

Bosco C, Colli R, Introini E, Cardinale M, Tsarpela O, Madella A, Tihanyi J, Viru A (1999b) Adaptive responses of human skeletal muscle to vibration exposure. Clin Physiol 19:183–187

Brislin RP, Theroux MC (2013) Core myopathies and malignant hyperthermia susceptibility: a review. Paediatr Anaesth 23:834–841

Burke D, Hagbarth KE, Lofstedt L, Wallin BG (1976) The responses of human muscle spindle endings to vibration of non-contracting muscles. J Physiol 261:673–693

Cardinale M, Bosco C (2003) The use of vibration as an exercise intervention. Exerc Sport Sci Rev 31:3–7

Carson RG, Riek S, Mackey DC, Meichenbaum DP, Willms K, Forner M, Byblow WD (2004) Excitability changes in human forearm corticospinal projections and spinal reflex pathways during rhythmic voluntary movement of the opposite limb. J Physiol 560:929–940

Chen J, Billings SE, Nishimune H (2011) Calcium channels link the muscle-derived synapse organizer laminin beta2 to Bassoon and CAST/Erc2 to organize presynaptic active zones. J Neurosci 31:512–525

Chen J, Mizushige T, Nishimune H (2012) Active zone density is conserved during synaptic growth but impaired in aged mice. J Comp Neurol 520:434–452

Cochrane DJ (2011) The potential neural mechanisms of acute indirect vibration. J Sports Sci Med 10:19–30

Delecluse C, Roelants M, Verschueren S (2003) Strength increase after whole-body vibration compared with resistance training. Med Sci Sports Exerc 35:1033–1041

Delecluse C, Roelants M, Diels R, Koninckx E, Verschueren S (2005) Effects of whole body vibration training on muscle strength and sprint performance in sprint-trained athletes. Int J Sports Med 26:662–668

Deschenes MR, Judelson DA, Kraemer WJ, Meskaitis VJ, Volek JS, Nindl BC, Harman FS, Deaver DR (2000) Effects of resistance training on neuromuscular junction morphology. Muscle Nerve 23:1576–1581

Deschenes MR, Wilson MH, Kraemer WJ (2005) Neuromuscular adaptations to spaceflight are specific to postural muscles. Muscle Nerve 31:468–474

Deschenes MR, Tenny KA, Wilson MH (2006) Increased and decreased activity elicits specific morphological adaptations of the neuromuscular junction. Neuroscience 137:1277–1283

Fahim MA (1997) Endurance exercise modulates neuromuscular junction of C57BL/6NNia aging mice. J Appl Physiol (1985) 83:59–66

Feng W, Tu J, Pouliquin P, Cabrales E, Shen X, Dulhunty A, Worley PF, Allen PD, Pessah IN (2008) Dynamic regulation of ryanodine receptor type 1 (RyR1) channel activity by Homer 1. Cell Calcium 43:307–314

Fitts RH, Riley DR, Widrick JJ (2001) Functional and structural adaptations of skeletal muscle to microgravity. J Exp Biol 204:3201–3208

Fling BW, Christie A, Kamen G (2009) Motor unit synchronization in FDI and biceps brachii muscles of strength-trained males. J Electromyogr Kinesiol 19:800–809

Flucher BE, Franzini-Armstrong C (1996) Formation of junctions involved in excitation-contraction coupling in skeletal and cardiac muscle. Proc Natl Acad Sci U S A 93:8101–8106

Fornari MC, Kohn AF (2008) High frequency tendon reflexes in the human soleus muscle. Neurosci Lett 440:193–196

Frank T, Rutherford MA, Strenzke N, Neef A, Pangrsic T, Khimich D, Fejtova A, Gundelfinger ED, Liberman MC, Harke B, Bryan KE, Lee A, Egner A, Riedel D, Moser T (2010) Bassoon and the synaptic ribbon organize Ca(2) + channels and vesicles to add release sites and promote refilling. Neuron 68:724–738

Gabriel DA, Kamen G, Frost G (2006) Neural adaptations to resistive exercise: mechanisms and recommendations for training practices. Sports Med 36:133–149

Gaspersic R, Koritnik B, Crne-Finderle N, Sketelj J (1999) Acetylcholinesterase in the neuromuscular junction. Chem Biol Interact 119–120:301–308

Gordon PM, Liu D, Sartor MA, IglayReger HB, Pistilli EE, Gutmann L, Nader GA, Hoffman EP (2012) Resistance exercise training influences skeletal muscle immune activation: a microarray analysis. J Appl Physiol (1985) 112:443–453

Gotti C, Clementi F, Fornari A, Gaimarri A, Guiducci S, Manfredi I, Moretti M, Pedrazzi P, Pucci L, Zoli M (2009) Structural and functional diversity of native brain neuronal nicotinic receptors. Biochem Pharmacol 78:703–711

Griffin L, Garland SJ, Ivanova T, Gossen ER (2001) Muscle vibration sustains motor unit firing rate during submaximal isometric fatigue in humans. J Physiol 535:929–936

Hagbarth KE, Eklund G (1966) Tonic vibration reflexes (TVR) in spasticity. Brain Res 2:201–203

Hallermann S, Fejtova A, Schmidt H, Weyhersmuller A, Silver RA, Gundelfinger ED, Eilers J (2010) Bassoon speeds vesicle reloading at a central excitatory synapse. Neuron 68:710–723

Harpsoe K, Ahring PK, Christensen JK, Jensen ML, Peters D, Balle T (2011) Unraveling the high- and low-sensitivity agonist responses of nicotinic acetylcholine receptors. J Neurosci 31: 10759–10766

Hong W, Lev S (2014) Tethering the assembly of SNARE complexes. Trends Cell Biol 24:35–43

Huang G, Kim JY, Dehoff M, Mizuno Y, Kamm KE, Worley PF, Muallem S, Zeng W (2007) Ca^{2+} signaling in microdomains: Homer1 mediates the interaction between RyR2 and Cav1.2 to regulate excitation-contraction coupling. J Biol Chem 282:14283–14290

Huang GN, Huso DL, Bouyain S, Tu J, McCorkell KA, May MJ, Zhu Y, Lutz M, Collins S, Dehoff M, Kang S, Whartenby K, Powell J, Leahy D, Worley PF (2008) NFAT binding and regulation of T cell activation by the cytoplasmic scaffolding Homer proteins. Science 319:476–481

Hunter DD, Shah V, Merlie JP, Sanes JR (1989) A laminin-like adhesive protein concentrated in the synaptic cleft of the neuromuscular junction. Nature 338:229–234

Kanzleiter T, Rath M, Görgens SW, Jensen J, Tangen DS, Kolnes AJ, Kolnes KJ, Lee S, Eckel J, Schürmann A, Eckardt K (2014) The myokine decorin is regulated by contraction and involved in muscle hypertrophy. Biochem Biophys Res Commun 25:1089–1094

Kiyonaka S, Wakamori M, Miki T, Uriu Y, Nonaka M, Bito H, Beedle AM, Mori E, Hara Y, De Waard M, Kanagawa M, Itakura M, Takahashi M, Campbell KP, Mori Y (2007) RIM1 confers sustained activity and neurotransmitter vesicle anchoring to presynaptic Ca^{2+} channels. Nat Neurosci 10:691–701

Kramer A, Gollhofer A, Ritzmann R (2013) Acute exposure to microgravity does not influence the H-reflex with or without whole body vibration and does not cause vibration-specific changes in muscular activity. J Electromyogr Kinesiol 23:872–878

Luo C, Shaw KT, Raghavan A, Aramburu J, Garcia-Cozar F, Perrino BA, Hogan PG, Rao A (1996) Interaction of calcineurin with a domain of the transcription factor NFAT1 that controls nuclear import. Proc Natl Acad Sci U S A 93:8907–8912

Mancini A, Sirabella D, Zhang W, Yamazaki H, Shirao T, Krauss RS (2011) Regulation of myotube formation by the actin-binding factor drebrin. Skelet Muscle 1:36

Mester J, Kleinoder H, Yue Z (2006) Vibration training: benefits and risks. J Biomech 39: 1056–1065

Mischi M, Cardinale M (2009) The effects of a 28-Hz vibration on arm muscle activity during isometric exercise. Med Sci Sports Exerc 41:645–653

Mukherjee K, Yang X, Gerber SH, Kwon HB, Ho A, Castillo PE, Liu X, Sudhof TC (2010) Piccolo and bassoon maintain synaptic vesicle clustering without directly participating in vesicle exocytosis. Proc Natl Acad Sci U S A 107:6504–6509

Munte TF, Jobges EM, Wieringa BM, Klein S, Schubert M, Johannes S, Dengler R (1996) Human evoked potentials to long duration vibratory stimuli: role of muscle afferents. Neurosci Lett 216:163–166

Nishimune H, Sanes JR, Carlson SS (2004) A synaptic laminin-calcium channel interaction organizes active zones in motor nerve terminals. Nature 432:580–587

Nishimune H, Numata T, Chen J, Aoki Y, Wang Y, Starr MP, Mori Y, Stanford JA (2012) Active zone protein Bassoon co-localizes with presynaptic calcium channel, modifies channel function, and recovers from aging related loss by exercise. PLoS One 7:e38029

Nishimune H, Stanford JA, Mori Y (2014) Role of exercise in maintaining the integrity of the neuromuscular junction. Muscle Nerve 49:315–324

Person R, Kozhina G (1992) Tonic vibration reflex of human limb muscles: discharge pattern of motor units. J Electromyogr Kinesiol 2:1–9

Pietrini V, Marbini A, Galli L, Sorrentino V (2004) Adult onset multi/minicore myopathy associated with a mutation in the RYR1 gene. J Neurol 251:102–104

Riley DA, Ilyina-Kakueva EI, Ellis S, Bain JL, Slocum GR, Sedlak FR (1990) Skeletal muscle fiber, nerve, and blood vessel breakdown in space-flown rats. FASEB J 4:84–91

Ritzmann R, Kramer A, Bernhardt S, Gollhofer A (2014) Whole body vibration training – improving balance control and muscle endurance. PLoS One 9:e89905

Roche KW, Tu JC, Petralia RS, Xiao B, Wenthold RJ, Worley PF (1999) Homer 1b regulates the trafficking of group I metabotropic glutamate receptors. J Biol Chem 274:25953–25957

Roll JP, Vedel JP, Ribot E (1989) Alteration of proprioceptive messages induced by tendon vibration in man: a microneurographic study. Exp Brain Res 76:213–222

Salanova M, Priori G, Barone V, Intravaia E, Flucher B, Ciruela F, McIlhinney RA, Parys JB, Mikoshiba K, Sorrentino V (2002) Homer proteins and InsP(3) receptors co-localise in the longitudinal sarcoplasmic reticulum of skeletal muscle fibres. Cell Calcium 32:193–200

Salanova M, Bortoloso E, Schiffl G, Gutsmann M, Belavy DL, Felsenberg D, Furlan S, Volpe P, Blottner D (2011) Expression and regulation of Homer in human skeletal muscle during neuromuscular junction adaptation to disuse and exercise. FASEB J 25:4312–4325

Salanova M, Volpe P, Blottner D (2013) Homer protein family regulation in skeletal muscle and neuromuscular adaptation. IUBMB Life 65:769–776

Salanova M, Gelfi C, Moriggi M, Vasso M, Viganò A, Minafra L, Bonifacio G, Schiffl G, Gutsmann M, Felsenberg D, Cerretelli P, Blottner D (2014) Disuse deterioration of human skeletal muscle challenged by resistive exercise superimposed with vibration: evidence from structural and proteomic analysis. FASEB J 28(11):4748–4763. doi:10.1096/fj.14-252825, Epub 2014 Aug 13

Sanes JR, Lichtman JW (1999) Development of the vertebrate neuromuscular junction. Annu Rev Neurosci 22:389–442

Shiraishi-Yamaguchi Y, Furuichi T (2007) The Homer family proteins. Genome Biol 8:206

Sitja Rabert M, Rigau Comas D, Fort Vanmeerhaeghe A, Santoyo Medina C, Roque i Figuls M, Romero-Rodriguez D, Bonfill Cosp X (2012) Whole-body vibration training for patients with neurodegenerative disease. Cochrane Database Syst Rev 2, CD009097

Soderpalm AC, Kroksmark AK, Magnusson P, Karlsson J, Tulinius M, Swolin-Eide D (2013) Whole body vibration therapy in patients with Duchenne muscular dystrophy–a prospective observational study. J Musculoskelet Neuronal Interact 13:13–18

Staron RS, Karapondo DL, Kraemer WJ, Fry AC, Gordon SE, Falkel JE, Hagerman FC, Hikida RS (1994) Skeletal muscle adaptations during early phase of heavy-resistance training in men and women. J Appl Physiol (1985) 76:1247–1255

Stiber JA, Tabatabaei N, Hawkins AF, Hawke T, Worley PF, Williams RS, Rosenberg P (2005) Homer modulates NFAT-dependent signaling during muscle differentiation. Dev Biol 287:213–224

Stiber JA, Zhang ZS, Burch J, Eu JP, Zhang S, Truskey GA, Seth M, Yamaguchi N, Meissner G, Shah R, Worley PF, Williams RS, Rosenberg PB (2008) Mice lacking Homer 1 exhibit a skeletal myopathy characterized by abnormal transient receptor potential channel activity. Mol Cell Biol 28:2637–2647

Sun LW, Blottner D, Luan HQ, Salanova M, Wang C, Niu HJ, Felsenberg D, Fan YB (2013) Bone and muscle structure and quality preserved by active versus passive muscle exercise on a new stepper device in 21 days tail-suspended rats. J Musculoskelet Neuronal Interact 13:166–177

Tothova J, Blaauw B, Pallafacchina G, Rudolf R, Argentini C, Reggiani C, Schiaffino S (2006) NFATc1 nucleocytoplasmic shuttling is controlled by nerve activity in skeletal muscle. J Cell Sci 119:1604–1611

Uriu Y, Kiyonaka S, Miki T, Yagi M, Akiyama S, Mori E, Nakao A, Beedle AM, Campbell KP, Wakamori M, Mori Y (2010) Rab3-interacting molecule gamma isoforms lacking the Rab3-binding domain induce long lasting currents but block neurotransmitter vesicle anchoring in voltage-dependent P/Q-type Ca^{2+} channels. J Biol Chem 285:21750–21767

Wang CZ, Wang GJ, Ho ML, Wang YH, Yeh ML, Chen CH (2010) Low-magnitude vertical vibration enhances myotube formation in C2C12 myoblasts. J Appl Physiol (1985) 109:840–848

Wilson MH, Deschenes MR (2005) The neuromuscular junction: anatomical features and adaptations to various forms of increased, or decreased neuromuscular activity. Int J Neurosci 115:803–828

Xiao B, Tu JC, Petralia RS, Yuan JP, Doan A, Breder CD, Ruggiero A, Lanahan AA, Wenthold RJ, Worley PF (1998) Homer regulates the association of group 1 metabotropic glutamate receptors with multivalent complexes of homer-related, synaptic proteins. Neuron 21:707–716

Chapter 4
Physical Countermeasure in Space: Efforts in Vain?

Abstract This chapter gives a brief overview on the currently used physical countermeasures onboard the ISS. Based on recent skeletal muscle research data from our laboratory, we here provide a personal perspective on the still unknown answer how to prevent disused skeletal muscle fibers following gravitational unloading in microgravity from atrophy, for example, by using alternative modes of physical exercise as countermeasure in spaceflight. We still do not know why the outcome of current inflight exercise prescriptions was of little avail to the crew members, and, for example, if optimized, exercise modes targeted to cellular and molecular properties of human skeletal muscle might help to overcome disuse-induced atrophy and impaired performance control of crew members during their long spaceflight missions. Based on the exciting strengthening effects of frequency-controlled vibration mechanosignals on the key cell signaling pathways involved in the structure, function, and free radical stress management of skeletal muscle fibers and neuromuscular synaptic junctions, we propose RVE as an additional highly efficient, short duration, and compliant exercise regimen (with thousands of contraction cycles induced by neuroreflexive muscle stimulation within a few minutes of training bouts) for the safety and health of crew members in preflight training, during spaceflight missions, and during recovery on Earth.

Keywords Skeletal muscle • Neuromuscular system • Exercise countermeasure • Spaceflight • Resistive vibration exercise • Nitric oxide signaling • Homer

4.1 Current Inflight Countermeasures and Perspectives

In order to overcome or mitigate the known human musculoskeletal atrophy in microgravity (Fitts et al 2001), a set of various physical exercise inflight counter-measure protocols including treadmill and bicycle ergometer (e.g., CEVIS) for cardiovascular support (aerobic interval / continuous) and a set of resistive exercise devices (IRED, ARED) are currently used by the crew members onboard the ISS. Another interesting approach uses low negative body pressure (LBMP) in combination with aerobic exercise on a treadmill to counteract cardiovascular system deconditioning and orthostatic intolerance by a thoraco-cephalic body fluid shifting which is a well-known phenomenon observed in long-duration supine body position

© The Author(s) 2015

D. Blottner, M. Salanova, *The NeuroMuscular System: From Earth to Space Life Science*, SpringerBriefs in Space Life Sciences, DOI 10.1007/978-3-319-12298-4_4

on the ground in bed rest, in jet pilots, as well as in microgravity (Macias et al. 2005). Most of the exercise devices were tested on the ground for their feasibility and efficacy to mitigate loss in muscle mass and force in the bed rest spaceflight analogue (Alkner and Tesch 2004; Watenpaugh et al. 2007). So far, the physical training devices used onboard the ISS, for example, are equivalents of aerobic cardiovascular regimen and weightlifting that are effectively used in everyday health and fitness workout but also in athlete sports on Earth.

In microgravity, resistive-like exercise can be performed if the almost unloaded human body (loss of G-forces) is "reloaded," for example, by elastic bungee straps and comfortable body harnessing systems that press the body and legs toward a belt conveyor or to a fixed and/or moveable platform of an inflight exercise device to produce muscle loading G-forces during, e.g., squatting bouts via simulated gravitational forces. First results from inflight resistive exercise regimen on the ISS showed some unexpected insufficient effects for musculoskeletal system support (Gopalakrishnan et al. 2009; Trappe et al. 2009). Follow-up inflight studies determined the foot forces in order to find improved ways for a better countermeasure outcome (duration/day) with increased body loading for the crew members requiring greater loadings during onboard exercise prescriptions (Genc et al. 2010), but inflight force measurements revealed that astronauts could not get even close to the pull of gravity on Earth. For example, NASA astronauts spent hours a day on the ISS for exercise on treadmills and other devices to combat muscle wasting and bone loss, however, with little avail (Witze 2014).

One possible explanation for this critical situation is given by the fact that the outcome of a strenuous resistive exercise regimen with high muscle tension, perfusion rate and neuromuscular recruitment, and maximally force output that works fine to maintain or even to build-up skeletal muscle performed under normal terrestrial G-force loading however is significantly altered following body unloading in Space possibly due to more skeletal muscle relaxation with relatively low neuromuscular activity and due to the inertial mass acceleration in microgravity. In Space, human motions may be more or less comparable to the variable biomechanical dynamics during skeletal muscle loading with very little torque observed after passive rather than active mode motions performed on Earth as, for example, confirmed by differential outcome of active and passive muscle training on muscle and bone quality in simulated microgravity with HU animals on the ground (Sun et al. 2013). Another possible explanation for an insufficient outcome of currently available inflight countermeasure protocols is given by the likely possibility of maladapted proprioceptive postural control of muscle activity in the unloaded crew member's body in Space which may not be adequately addressed by increasing the resistive loading during exercise bouts, for example, due to altered kinematic synergies in complex movements in microgravity (Casellato et al. 2007).

The current knowledge on inflight countermeasure protocols suggests that more sophisticated countermeasure paradigms should be exploited taking into account the obvious different biomechanical and neuromuscular control mechanisms in human motion and performance on Earth and in Space. As the complex physiological mechanisms of gravitational force sensing, transmission, and support are not yet fully understood for most of the biological soft and hard tissues including their neural adaptation mechanisms in the microgravity environment, additional basic investigations are needed in this direction in order to learn more about the physiological gravisensor mechanisms of the normal and deconditioned neuromuscular system and the movement kinematics modulated by altered gravity (0–1G, ΔG). This would be important in order to find feasible ways of counter-measure protocols with new technology that is sufficiently compliant for crew members to be used in spaceflight in order to minimize at least some of their health risks onboard the ISS until 2024 (Witze 2014), and possibly thereafter, in particular using more adequate inflight countermeasure prescriptions are required for the benefit of the next generation of space travelers during their prolonged space missions to Moon and Mars.

Our recent work and work by other laboratories have accumulated a set of compelling evidence in animal as well as human studies using multidisciplinary up-to-date analysis to show that if compared to standard RE protocols currently used as inflight countermeasure in spaceflight, the short but effective RVE stimu-lation and the mechanosignals and neurosignals elicited are targeted to a number of key signaling molecules in muscle fibers and at the neuromuscular junction (NOS1, PGC1a, NFkB, Homer). Even more, the power and force-associated functional muscle protein expressions at sarcomeres (the little power chambers in myofibers) including their ultrastructure integrity are supported by RVE as compared to conventional resistive exercise countermeasure. In addition to the obvious preser-vation of the muscle size and myofiber phenotype distribution by both RE and RVE, most of the muscle-specific signaling pathways and their related structural and contractile proteins investigated are particularly downregulated in human muscle disuse and are upregulated in particular by RVE (Blottner et al 2006; Salanova et al. 2008; Salanova et al. 2009; Salanova et al 2011; Salanova et al. 2013a; Salanova et al. 2013b; Moriggi et al 2010; Salanova et al. 2014). The present findings are in favor of our hypothesis that, compared to standard RE, muscle training by RVE has additional positive effects on key molecular pathways and functional proteins involved in disuse atrophy and should likely serve as an alternative therapeutic exercise regimen to combat atrophy in longer periods of skeletal muscle disuse when used on the ground, for example, in rehabilitation or various other clinical settings. We here propose RVE to be implemented as additional protocol to support adequate neuromuscular system control for the safety and health of the crew members in preflight training, during spaceflight missions, and during recovery on Earth.

References

Alkner BA, Tesch PA (2004) Knee extensor and plantar flexor muscle size and function following 90 days of bed rest with or without resistance exercise. Eur J Appl Physiol 93:294–305

Blottner D, Salanova M, Püttmann B, Schiffl G, Felsenberg D, Buehring B, Rittweger J (2006) Human skeletal muscle structure and function preserved by vibration muscle exercise following 55 days of bed rest. Eur J Appl Physiol 97:261–271

Casellato C, Tagliabue M, Pedrocchi A, Ferrigno G, Pozzo T (2007) How does microgravity affect the muscular and kinematic synergies in a complex movement? J Gravit Physiol 14(1):93–94

Fitts RH, Riley DR, Widrick JJ (2001) Functional and structural adaptations of skeletal muscle to microgravity. J Exp Biol 204:3201–3208

Genc KO, Gopalakrishnan R, Kuklis MM, Maender CC, Rice AJ, Bowersox KD, Cavanagh PR (2010) Foot forces during exercise on the International Space Station. J Biomech 43: 3020–3027

Gopalakrishnan R, Genc KO, Rice AJ, Lee SM, Evans HJ, Maender CC, Ilaslan H, Cavanagh PR (2009) Muscle volume, strength, endurance, and exercise loads during 6-month missions in space. Aviat Space Environ Med 81:91–102

Macias BR, Groppo ER, Eastlack RK, Watenpaugh DE, Lee SM, Schneider SM, Boda WL, Smith SM, Cutuk A, Pedowitz RA, Meyer RS, Hargens AR (2005) Space exercise and Earth benefits. Curr Pharm Biotechnol 6:305–317. LBMP as countermeasure

Moriggi M, Vasso M, Fania C, Capitanio D, Bonifacio G, Salanova M, Blottner D, Rittweger J, Felsenberg D, Cerretelli P, Gelfi C (2010) Long term bed rest with and without vibration exercise countermeasures: effects on human muscle protein dysregulation. Proteomics 10: 3756–3774

Salanova M, Schiffl G, Püttmann B, Schoser BG, Blottner D (2008) Molecular biomarkers monitoring human skeletal muscle fibres and microvasculature following long-term bed rest with and without countermeasures. J Anat 212:306–318

Salanova M, Schiffl G, Rittweger J, Felsenberg D, Blottner D (2009) Ryanodine receptor type-1 (RyR1) expression and protein S-nitrosylation pattern in human soleus myofibres following bed rest and exercise countermeasure. Histochem Cell Biol 130:105–118

Salanova M, Bortoloso E, Schiffl G, Gutsmann M, Belavy DL, Felsenberg D, Furlan S, Volpe P, Blottner D (2011) Expression and regulation of Homer in human skeletal muscle during neuromuscular junction adaptation to disuse and exercise. FASEB J 25:4312–4325

Salanova M, Volpe P, Blottner D (2013a) Homer protein family regulation in skeletal muscle and neuromuscular adaptation. IUBMB Life 65:769–776

Salanova M, Schiffl G, Gutsmann M, Felsenberg D, Furlan S, Volpe P, Clarke A, Blottner D (2013b) Nitrosative stress in human skeletal muscle attenuated by exercise countermeasure after chronic disuse. Redox Biol 1:514–526. doi:10.1016/j.redox.2013.10.006

Salanova M, Gelfi C, Moriggi M, Vasso M, Viganò A, Minafra L, Bonifacio G, Schiffl G, Gutsmann M, Felsenberg D, Cerretelli P, Blottner D (2014) Disuse deterioration of human skeletal muscle challenged by resistive exercise superimposed with vibration: Evidence from structural and proteomic analysis. FASEB J 28: 4748–4763

Sun LW, Blottner D, Luan HQ, Salanova M, Wang C, Niu HJ, Felsenberg D, Fan YB (2013) Bone and muscle structure and quality preserved by active versus passive muscle exercise on a new stepper device in 21 days tail-suspended rats. J Musculoskelet Neuronal Interact 13:166–177

Trappe S, Costill D, Gallagher P, Creer A, Peters JR, Evans H, Riley DA, Fitts RH (2009) Exercise in space: human skeletal muscle after 6 months aboard the International Space Station. J Appl Physiol 106:1159–1168

Watenpaugh DE, O'Leary DD, Schneider SM, Lee SM, Macias BR, Tanaka K, Hughson RL, Hargens AR (2007) Lower body negative pressure exercise plus brief postexercise lower body negative pressure improve post-bed rest orthostatic tolerance. J Appl Physiol 103(6): 1964–1972

Witze A (2014) Space-station science ramps up. Nature 510(7504):196–197. doi:10.1038/510196a